A Billion Little Pieces

Infrastructures Series

edited by Geoffrey C. Bowker and Paul N. Edwards

Paul N. Edwards, *A Vast Machine: Computer Models, Climate Data, and the Politics of Global Warming*

Lawrence M. Busch, *Standards: Recipes for Reality*

Lisa Gitelman, ed., *"Raw Data" Is an Oxymoron*

Finn Brunton, *Spam: A Shadow History of the Internet*

Nil Disco and Eda Kranakis, eds., *Cosmopolitan Commons: Sharing Resources and Risks across Borders*

Casper Bruun Jensen and Brit Ross Winthereik, *Monitoring Movements in Development Aid: Recursive Partnerships and Infrastructures*

James Leach and Lee Wilson, eds., *Subversion, Conversion, Development: Cross-Cultural Knowledge Exchange and the Politics of Design*

Olga Kuchinskaya, *The Politics of Invisibility: Public Knowledge about Radiation Health Effects after Chernobyl*

Ashley Carse, *Beyond the Big Ditch: Politics, Ecology, and Infrastructure at the Panama Canal*

Alexander Klose, translated by Charles Marcrum II, *The Container Principle: How a Box Changes the Way We Think*

Eric T. Meyer and Ralph Schroeder, *Knowledge Machines: Digital Transformations of the Sciences and Humanities*

Geoffrey C. Bowker, Stefan Timmermans, Adele E. Clarke, and Ellen Balka, eds., *Boundary Objects and Beyond: Working with Leigh Star*

Clifford Siskin, *System: The Shaping of Modern Knowledge*

Lawrence Busch, *Knowledge for Sale: The Neoliberal Takeover of Higher Education*

Bill Maurer and Lana Swartz, *Paid: Tales of Dongles, Checks, and Other Money Stuff*

Katayoun Shafiee, *Machineries of Oil: An Infrastructural History of BP in Iran*

Megan Finn, *Documenting Aftermath: Information Infrastructures in the Wake of Disasters*

Ann M. Pendleton-Jullian and John Seely Brown, *Design Unbound: Designing for Emergence in a White Water World*, Volume 1: *Designing for Emergence*

Ann M. Pendleton-Jullian and John Seely Brown, *Design Unbound: Designing for Emergence in a White Water World*, Volume 2: *Ecologies of Change*

Jordan Frith, *A Billion Little Pieces: RFID and Infrastructures of Identification*

A Billion Little Pieces

RFID and Infrastructures of Identification

Jordan Frith

The MIT Press
Cambridge, Massachusetts
London, England

This book was set in Stone Serif by Westchester Publishing Services.

Library of Congress Cataloging-in-Publication Data

Names: Frith, Jordan, author.
Title: A billion little pieces : RFID and the infrastructures of
  identification / Jordan Frith.
Other titles: RFID and the infrastructures of identification
Description: Cambridge, MA : The MIT Press, [2019] | Includes
  bibliographical references and index.
Identifiers: LCCN 2018033822 | ISBN 9780262039758 (hardcover :
  alk. paper), 9780262551281 (pb)
Subjects: LCSH: Radio frequency identification systems.
Classification: LCC TK6570.I34 F75 2019 | DDC 621.3841/92—dc23
  LC record available at https://lccn.loc.gov/2018033822

don't shop!). They kept me company as I wrote this book, and while I know they can't read, I trust that they'd be super interested in what I was doing if they could. After all, they're the ones who actually have RFID chips in their backs, and I became interested in RFID in part because I became interested in those identification microchips. So, yeah ... dogs are the best.

1 RFID and the Infrastructural Imagination

The world has become more and more mobile. More people travel across national borders. Products travel farther and more frequently in the global economy. Roads handle an increasing number of drivers. Mobility infrastructure is used more heavily than ever before. At the same time, people have also become more interested in data of all types and tracking mobility in new ways. We live in what has been called an "information society" that seeks to record everything it can.[1]

Lying at the convergence of these two trends—an increase of mobility and a growing focus on datafication—is a mobile technology relatively few people know about: radio-frequency identification (RFID). RFID is simultaneously a communicative mobile technology and an important part of various infrastructures. RFID systems include a tag that carries data, a reader that accesses that data, and middleware that transmits the data from the reader to a larger system. RFID tags can be as small as a grain of rice and can be sewn into clothing or embedded in packaging. They are found in credit cards, passports, key fobs, car windshields, subway cards, consumer electronics, the walls of tunnels, and even inside animal and human bodies.

So what is RFID? I explain the technical side in more detail in chapter 3, but for now think of RFID as a suite of technologies that uses radio waves to communicate identifying information. Many tags feature an Electronic Product Code (EPC)[2] that helps uniquely identify billions of tags in circulation. Other tags, such as those found in public transportation "smart" cards and contactless credit cards, transmit serial numbers that grant access to database information.[3] In other cases, RFID tags paired with sensor technology transmit environmental information, such as the temperature of food products or the wear on infrastructure.[4]

Many people reading this book are likely within reach of an RFID tag. There might be an antenna in their contactless credit card or a chip in their pet. If they checked this book out from the library, an RFID tag might be on the inside cover of the binding. Or readers may have used their smartphone to pay for something using a mobile payment application, which uses a modified form of RFID called near-field communication (NFC). But unless people look closely or start thinking about the implications of infrastructure, most of these tags will remain out of view, even as RFID technology has the potential to invisibly communicate data about the identity and movement of people and objects. One of the main goals of this book is to pull back the curtain on RFID, examining where to find the technology, the social impact of RFID as mobile identification infrastructure, and how RFID is one small technology that nonetheless showcases the larger push toward the Internet of Things and big data.

The question remains why anyone should read a book devoted to RFID, especially a book that looks at RFID from a sociological perspective. I believe that answer lies in what other scholars have identified as somewhat of a blind spot in some media studies research: not enough research has focused beneath the level of the interface.[5] Researchers tend to pay attention to the technologies they can see, the technologies with which people interact every day. More words than could possibly be discussed here have been written about the primary media technologies people use to communicate, ranging from the written word to new smartphone applications. People tend to ignore what happens in the background of their interactions with the physical world. That lack of attention is designed into the system. As sociologist Susan Leigh Star has argued, and countless articles have repeated, infrastructures are "by definition invisible" and they only "become visible upon breakdown."[6]

People notice infrastructure mostly when it stops working, not when it works as planned. Take mobile phone infrastructure as an example. Most people do not pay attention to cell phone towers, and engineers go to great lengths to hide towers by disguising them as buildings and trees.[7] Many people think about cell towers only when their coverage fails and they find themselves wandering around trying to pick up a signal. At its core, infrastructure is that which people rely on but rarely think critically about.[8]

What happens when infrastructure is ignored, however, is that researchers end up focusing only on the level of the interface. They emphasize the human intentionality behind people's communicative practices without acknowledging how much communication now occurs beneath the level of human perception.[9] Additionally, when people ignore established infrastructures, the infrastructures' histories are erased, and they become more invisible and seemingly magical than they really are. After all, as I explore in this book, infrastructures are not quite as invisible as many people think. Rather, infrastructures can be extremely visible and invisible in alternating patterns. An infrastructure can initially be met with widespread hype and marketed as transformative, making the technology anything but invisible. Once adopted and put in place, however, that hype fades, and the "cutting-edge" technology fades into the background. As anthropologist Brian Larkin put it, "Invisibility is certainly one aspect of infrastructure, but it is only one and at the extreme edge of a range of visibilities that move from unseen to grand spectacles and everything in between."[10]

RFID is a technology that exemplifies the range of visibilities Larkin discussed. On the one hand, the technology often fades into the background, and many people interact regularly with RFID technologies without knowing at all what they are. On the other hand, RFID has been a massively hyped and feared technology: the original concept of the "Internet of Things" focused on RFID, RFID has been identified as possibly the most important identification technology since the barcode, NFC (a type of RFID) was positioned as a technology that would alter how people interact with the built environment, and RFID has been met with widespread fears about surveillance in some communities. Some groups of evangelical Christians have even identified the technology as a possible harbinger of the end times. Consequently, it is not quite accurate to say that RFID is an "invisible" infrastructural technology. Throughout most of this book, I look at how RFID is embedded in objects and designed to be not noticed, but I also take into account the moments wherein RFID becomes hyper-visible, whether through the marketing of new capabilities or the fears about what the technology can tell larger entities about how objects exist in the world.

The hype and fears almost all center on the main technological capability of RFID: the ability to collect unique identification data about almost any material object. Consequently, much of this book focuses on what I call *object communication,* which I define as *the ability of objects that have either no or little computing power to wirelessly communicate identifying data with infrastructure.* Object communication is less complex than the more complicated types of machine-to-machine communication people are working on in the realms of artificial intelligence, but examples such as the ability of a pair of jeans with an RFID tag to communicate identification data wirelessly, a subway card to link contactlessly with a terminal, or a car windshield to be read while traveling full speed down a highway show how objects that have no internal computing power play an important communicative role in the functioning of physical spaces. In addition, each example involves a sorting of some type. After all, as I explain throughout the following chapters, many of the types of object communication I examine feed into larger infrastructures of identification. Through tags of various shapes and sizes, billions upon billions of objects and bodies can be differentiated from one another and feed into a large-scale technological scheme of classification and identification.

My argument in this book is that—with the growth of communicative objects—we need to pay more attention to infrastructure, but not just in the traditional understanding of media infrastructure as that which supports interpersonal communication. We need to think of infrastructure as *that which communicates.* We also need to understand the role various infrastructures play in uniquely identifying and sorting objects in the physical world. Consequently, I position RFID as an infrastructure of identification that works simultaneously as an infrastructure of communication. I show that the physical world is already communicative, and much of what it has to say involves telling the built environment exactly what this "thing" is that is moving through the world at a given point at a given time. I embark on that exploration of infrastructures of identification by using the technology of RFID as a jumping-off point to examine the links between mobile technology and infrastructure, focusing on one specific technology as a way to explore the changing landscape of object communication.[11]

Ultimately, this book is about RFID as mobile technology, but at the same time it is not just about technology. The chapters that follow feature explanations of RFID systems, but the goal is to do more than just write about a single type of technology. Rather, I use RFID to examine larger issues such as the Internet of Things, big data, and privacy and surveillance. Looking at tags and microchips may seem like a strange way to get at some of the biggest buzzwords in contemporary society, but limiting my focus to RFID provides concreteness to what often are vague discussions about societal trends. The focus also enables the chapters to develop the core argument of this book: *RFID is a communicative mobile technology that can uniquely identify and sort billions of objects and turn various physical processes into trackable data.* Everything from the pedigree of the prescription drug I fill to the identities of my rescue dogs can be communicated through RFID technology.

The chapters that follow tell stories of RFID technology, but they do so without falling back on dry, technical descriptions. Instead, I follow RFID from pilgrimages in India to the Book of Revelation. I look at how RFID can surveille handwashing in hospitals and shift the visibility of supply chains. I mention projects that use RFID to communicate wine vintages and groups that believe RFID is part of a vast governmental conspiracy. Most importantly, I use these examples to show how the ubiquity of RFID can give voice to objects in novel ways and provide the identification necessary for a communicative, differentiated environment of networked things. Before moving on to those examples, however, the rest of this chapter explains why RFID is important; I also situate my examination of the technology within existing research. I conclude the chapter by outlining the structure of the book and providing a roadmap that will take us on a journey from the tools in Australian hospitals to the restaurants at Disney World.

RFID as Mobile Media

Back in May 2014, I was giving a conference talk about the location-based mobile application Foursquare that enabled people to share their location with friends. The talk went well, and the audience seemed interested in what my panel had to say about the social practices of location sharing. In the Q&A, however, Rich Ling—one of the major figures in the field of mobile communication studies—asked a provocative question: Why do mobile media scholars study less-popular locative technologies rather than technologies like RFID? The point he was making is a simple one to anyone aware of both popular press and academic discussions of mobile locative media. Researchers and journalists tend to focus on the snazzy new smartphone app people use to find their location or look up information about their surroundings. The tech press, for example, wrote extensively on the "check-in wars" between applications like Foursquare and Gowalla,[12] though neither application passed fifty million users, and Gowalla had failed by 2012. Academic mobile media literature is filled with accounts of location-based gaming,[13] location-based social networking,[14] mobile mapping,[15] and location-based search.[16]

Not coincidentally, the forms of locative media covered extensively in the press and in sociological research focus mostly on location-based applications people use to learn about their surroundings. That focus makes sense. The buzz surrounding a cool new locative application is likely higher than with something like RFID tags that track people and things. A new mobile app is flashier than a supply-chain technology or a card reader in a train station. And I do not point out the abundance of locative media research as a criticism. Much of the research is excellent and important. My first two books examine smartphone technology,[17] so I understand why people pay more attention to location-based mobile games that never approached a million users than they do to RFID technology that has been deployed in the billions. Media scholars are accustomed to focusing on user practices, and mobile applications—at the level in which someone interacts with an interface—are where those user practices are typically found.

But returning to Rich Ling's question, the fact remains that relatively ignored technologies like RFID tags are used more frequently to track information about people and things than are mobile applications. Many of the locative applications or locative artworks that have been extensively researched never surpass a few million users, and that is the high end in many cases. RFID tags, on the other hand, have been deployed in the tens of billions. One RFID company—Impinj—has put more than ten billion tags into circulation and expects to have over one hundred billion tags in circulation by 2020.[18] And that is only one company. The actual number of RFID tags deployed is impossible to calculate definitively, but the forecasting firm IDTechEx estimates that, in 2017, 8.7 billion RFID labels were used just in retail, another 825 million tags were used in transportation, and

480 million tags were used to meet animal tagging regulations.[19] In total, the RFID industry is worth more than $11 billion USD and will almost certainly continue to grow.

Of course, popularity does not equal importance. Billions of staples have been used in the last hundred years as a mobile technology, but no one is writing a book about staples (though I would probably read it). Additionally, the fact that not much sociological research exists on RFID does not automatically mean there *should* be research on the topic. So why look specifically at RFID? Mainly because RFID forms crucial infrastructures of identification that shape our world in diverse ways, enabling everything from certain types of physical movement to new retail practices powered through data. The technology moves with objects and bodies and links them with the built environment, most often in a process that involves individuating one object from others of its type. Consequently, RFID can show how increasingly distinct identification practices rely on material infrastructure to govern how bodies and objects move through networks of transportation and supply.

Much has been written in science and technology studies (STS) research about how classification systems work to differentiate people and things, make them identifiable, and place them in categories. Sociologists Geoffrey C. Bowker and Susan Leigh Star, for example, have written extensively about how infrastructures of classification work to "sort things out," whether those infrastructures involve disease definitions or racial sorting.[20] Less has been written, however, on the types of sorting enabled by mobile technologies like barcodes and RFID. As explored throughout these chapters, the increased data capacity of RFID can sort billions of objects from one another. These infrastructures of identification have significant implications for the communicative networks that shape our physical spaces, the relationships between bodies and institutions, and the increasingly detailed forms of identification necessary for the Internet of Things. While RFID is only one example of a technology used to semi-automate the process of "sorting things out," it is a crucial example whose story remains untold and should be an important piece of the mobile media landscape.

In addition, an argument that underlies the examples in this book is that RFID is representative of a larger trend toward more and more granular practices of identification. In chapter 2, I look at how the move from barcodes to RFID is a move toward enhanced specificity. That move is part of the larger trend—a trend that includes everything from biometric scanners to enhanced driver's licenses to improved analytic techniques for analyzing web traffic—of practices enabled through hardware and software that sort and differentiate one body or one thing from all the others. Consequently, while the focus of this book remains on RFID, the shifts in specificity of data collection are also likely just as relevant to various mass collections of data that shape our lives. This book is about separating and segmenting the individual (defined broadly) from the mass. That segmentation may involve differentiating one cow from all the others or one student in a classroom from her classmates. Regardless, the practices rely on the tiny tags and the data they produce to provide new types of visibility about movement and identification in the physical world.

The uses of RFID for producing the types of data discussed above are diverse. The data on tags may relate to someone's mobility in the case of toll tags or subway cards; they may relate to supply-chain logistics or the inventory on store shelves; or they may relate to the state of any object communicating with another object in the Internet of Things. But in all these cases, attaching or embedding RFID tags in objects turns the objects into animated, communicative, and uniquely identifiable pieces of the physical environment. Often the tags themselves do not even have an internal power source, so they can cost as little as ten cents a tag and work in near perpetuity because they have no battery to be replaced. The tags can then make literally tens of billions of objects uniquely traceable. They help enact a form of mobile communication that can do anything from open a toll-gate to log an item's location in a blockchain. By no means should scholars abandon the focus on interpersonal mobile communication, but I hope my examination of RFID will join other infrastructural research in expanding analyses of mobile media to look beneath the level of direct engagement with mobile interfaces.

Additionally, while this book is about RFID technology, it is not *about* RFID in the traditional sense of just describing the current state of a technology. Instead, as I hope to make clear in proceeding chapters, RFID is interesting not just because of its technological features but because it works as a window through which to explore larger societal issues. The examples I describe show how one small, often ignored, technology works in the background to control, standardize, and enable data projects, the Internet of Things, and corporeal mobility. I build on the work of Bowker, Star, and others to explore the increasingly complex practices used to sort, differentiate, and classify objects and bodies in the physical world. More than just expanding on processes of "sorting things out," however, I use RFID as a way to show the increasingly crucial role communicative technologies play in processes of identification, aiding and building on the various classification systems that already exist. Additionally, one of the key contributions of this book is to show how RFID represents a shift in the increasingly fine-grained processes of sorting and differentiation. Identification of various types has only become a more important piece of the "technological unconscious" that shapes our world, and RFID is a specific site at which to track that shift.

Consequently, readers coming to this book for a how-to guide for RFID deployment or a full account of the contemporary RFID landscape will likely be disappointed. I focus on the technology, but this book is not just about technology. If this book is successful, then the points I make, the issues I explore, will be relevant to people interested in infrastructures of identification and people interested in more traditional mobile communication. The topics I discuss in the forthcoming chapters will build on other works that have examined how a variety of sociotechnical assemblages, ranging from algorithms to databases to physical infrastructure, have reshaped what it means to communicate in the contemporary world.

Contextualizing RFID

A broad claim that RFID is an understudied technology is a bit of an overstatement. Technical and business-oriented fields feature hundreds of articles about RFID use. Computer science journals have published many articles about using RFID to interconnect networked devices.[21] Engineering journals have included articles about using RFID and sensors to monitor networked urban infrastructure.[22] Marketing journals have featured articles about using near-field communication to connect with customers.[23] Logistics publications have included articles and books about using RFID to track products and manage inventory.[24] The RFID industry also has the *RFID Journal*, which is a trade publication focused on RFID-related issues.

RFID has received less attention, on the other hand, from the social sciences and humanities. A few important discussions of "software-sorted geographies,"[25] "code/spaces,"[26] and smart city surveillance[27] mention RFID, but only as one small part of a more broadly defined socio-technical assemblage. Three significant exceptions to the lack of research from social scientific and humanistic perspectives were articles by Nigel Thrift,[28] Katherine Hayles,[29] and Martin Dodge and Rob Kitchin[30] in the mid-2000s. Thrift identified RFID as a key piece of the "technological unconscious" that operates in the background of people's everyday interactions with their surrounding spaces. He wrote that "the fourth innovation, and perhaps in the end the one likely to prove the most powerful, is the RFID (Radio Frequency Identification) tag."[31] Yet despite more than two hundred citations to his article, few scholars took up his provocation to analyze and theorize the impacts of RFID technology. One exception was media scholar Katherine Hayles, who argued that the central theoretical concerns about the technology "are the effects of RFID in creating an animate environment with agential and communicative powers."[32] She also provoked researchers to develop a framework that bridges human agency and RFID without collapsing distinctions between the two. Such a framework would allow us to "shed the misconception that humans alone are capable of cognition."[33] Finally, geographers Martin Dodge and Rob Kitchin also wrote explicitly about RFID, using the technology as an example of how the "application and automatic processing of digital identification codes are key to the evolving forms of contemporary governmentality and capitalism."[34]

I return to these articles throughout this book, but for now I want to point out that—despite these thinkers' status as prominent figures in the study of technology and society—few have stepped forward to address their provocations. The researchers cited above identified RFID as a potentially transformative technology that could reshape processes of identification and the composition of physical spaces, but few articles have taken RFID on its own terms and analyzed its impact on space and place, agency, or surveillance. The technology has remained in the background of larger discussions of infrastructure, urban studies, and global flows of products and people. This book shifts the foreground/background relationship by moving RFID to the forefront of discussions of mobile technology. Or, in Bowker and Star's terms, part of the goal of this book is to perform an "infrastructural inversion" that focuses explicitly on infrastructure.[35] Throughout these chapters, I examine how RFID as infrastructure of identification can alter relationships between bodies and institutions, between objects and databases, and among communicative material "things" that become networked in new ways. After all, with the hype surrounding the Internet of Things, it will become increasingly important to engage with the effects of the hardware and software that make it possible to connect and identify objects. That engagement is even more important because, as I examine next, much of the language people use to discuss the internet erases the physical infrastructures that make new communication practices possible.

Determinism and the Internet of Things

When people talk about the internet, they often slip into imagining the internet as a space somehow separate from the physical world. Theories of the internet often imagined a "cyberspace" separate from physical space. The same is true for other digital technologies. For instance, mobile phone researchers talked about forms of "absent presence" in which people engage with the digital spaces of the phone call and ignore their physical surroundings.[36] By no means is the tendency to separate the digital and physical confined to academic arguments. The popular press loves articles about how smartphones distract people or how internet use stops people from engaging with people or places nearby. Even popular metaphors like "The Cloud" suggest a separation between the physical and digital by implying the digital is some kind of ethereal substance that does not rely on tangible, material infrastructure.

Despite the tendency to dichotomize the physical and the digital, the digital is of course always physical. Digital information is not stored on a cloud somewhere; it is not "virtual" as opposed to "real." But with the growth of the Internet of Things, the idea that the digital and the physical could ever be separated has become even more tenuous, if not outright impossible, to maintain. The Internet of Things refers to objects that can connect with each other and with the internet. In other words, everyday objects become networked and shape the way data about the physical world is produced and collected. The term has become a marketing buzzword, with sources predicting that tens of billions of objects will be connected to the internet and to each other.[37] But through all the hype, the concept gets at a core, straightforward idea: objects will increasingly form networked connections and have new voices with which to speak and share data.

As I explore in more detail in chapter 4, according to some sources, objects in the Internet of Things must have their own internet connection. However, the original concept of the Internet of Things was introduced in a 1999 presentation about RFID technology,[38] and if tens of billions of objects are connected in the Internet of Things, many of those objects will be too small and too cheap to have their own internet connection. An RFID tag can make objects machine-readable, uniquely identifiable, and communicative. The tag can enroll anything from a pair of pants to a child walking through Disney World into a communicative infrastructural network. Possibly most important, typical RFID tags have the storage capacity to assign a unique identifying number to hundreds of billions of objects. Consequently, the identification and communication affordances of RFID are why the technology, in all the different forms I explore in these chapters, is a key piece of the growing Internet of Things.

Merely pointing out that RFID is a part of the Internet of Things misses a larger point about the agential effects of the technology. In her article about RFID, Katherine Hayles's provocation was that researchers needed to build a bridge between human agency and the agency of RFID. Doing so would necessitate avoiding the two poles that often dominate discussions of technologies: technological determinism and cultural determinism. On the one hand, RFID does not come from nowhere and then fully determine behavior in any technologically deterministic sense. The fact that radio waves exist and have agency does not mean it was inevitable that schools in Brazil would find a way to track students with RFID tags. On the other hand, RFID does have agential effects that allow certain behaviors and disallow others. The technology makes certain ways of acting and certain processes of identification possible. Without the technology, certain behaviors and forms of identification data just would not exist.

Consequently, conceptualizing the agency of RFID requires theories that avoid the tendency toward determinism. In the space between cultural and technological determinism, social theories have arisen that recognize the agency of technology. Theoretical movements such as object-oriented ontology,[39] activity theory,[40] phenomenology,[41] assemblage theory,[42] vital materialism,[43] and actor-network theory[44] all complicate questions of agency, distributing the ability to act and influence among networks of humans and things. These theoretical foundations all have major differences, but what they share might be more important: a recognition that things do have agency and are worth studying on their own terms.

These theories of nonhuman agency shape the examples I analyze in this book and how I talk about RFID's role as an infrastructure of identification. The foundation of my approach, an approach that argues that RFID technology is an active agent in the datafication of the world, is based on the work of people like Bruno Latour,[45] Jane Bennett,[46] John Law,[47] and many others who argue for distributed approaches to understanding action. I view RFID as an actant, defined as "that which has efficacy, can do things, has sufficient coherence to make a difference, produce effects, alter the course of events."[48] As I show throughout this book, RFID has the potential to alter the course of events and contribute to new forms of data collection and new types of object communication. I am not arguing that RFID causes larger societal movements like big data or the Internet of Things, but I also do not believe RFID merely responds to the push toward massive increases in volumes of data. Instead, RFID's role is both shaped by existing trends and helps shape those trends. The technology acts in concert (or not in concert, in some cases) with diverse and widespread networks of both humans and things.

Recognition of technological agency does not mean that this book is just about a technology. The stories I tell feature humans and animals, not just radio waves and antennae. My arguments are informed by the people I interviewed, the lab directors I followed, and the trade shows I attended. Human actors play a role in this book, as they should when discussing technology. After all, RFID does not come from outside society. But humans are not always front and center throughout this book because I am equally interested in exploring the agency of the technology. In effect, drawing on actor-network theory, I do not view technology and society as separate.[49] The actions of humans and things work together to hold what people think of as society together. The recognition of a flatter ontology of agency shows the value of tracing RFID tags through the environment. After all, as political scientist Jane Bennett writes, there is "public value in following the scent of a nonhuman, thingly power, the material agency of natural bodies and technological artifacts."[50]

Objects and Their Stories

A major part of the Internet of Things, especially in relation to RFID, involves tracking mobility. RFID tags, after all, can be thought of as a form of locative media.[51] Despite the warnings of some conspiracy theorists I discuss later, the tags do not use locative functions such as GPS or Wi-Fi triangulation, but they do transmit identification data about where an object is at a given time because they transmit to fixed readers. Consequently, the system often knows where the reader is, so it knows where the tag is. As one example, an RFID-enabled school-attendance system shows where and when a student enters a classroom because their RFID device wirelessly connects with a specifically located reader. Consequently, RFID is an important infrastructural technology in what has been called the "mobilities turn,"[52] which is a movement in the social sciences and humanities that focuses on the importance of mobility in shaping contemporary life. As Mimi Sheller argued, one of the keys to mobilities research is the ability to analyze the effects of "infrastructures of mobility in creating the effects of both movement and stasis."[53]

The mobilities turn was based partially on the work of geographers such as Doreen Massey, who argued that place cannot be understood without understanding flows of people, objects, and information.[54] Following Massey and other thinkers, sociologists Mimi Sheller and John Urry argued that "travel has been for the social sciences seen as a black box, a neutral set of technologies and processes."[55] One of the goals of mobilities research is to open up that black box to better understand how objects, people, and information flow in the globally networked world.

As both a mobile technology and a piece of communicative identification infrastructure, RFID can create logs of objects' mobility. The technology can, in other words, help objects tell their—admittedly truncated—stories. The objects can be turned into lesser versions of what futurist Bruce Sterling called "spimes."[56] Sterling argued that society is heading toward a new age of technological development in which objects will become interactive and able to communicate their stories with humans. He defines spimes as "manufactured objects whose information support is so overwhelmingly extensive and rich that they are regarded as material instantiations of an immaterial system. Spimes begin and end as data.... Spimes are sustainable, enhanceable, uniquely identifiable."[57]

As wide-ranging as that definition may seem, Sterling is describing a rather clear future in which objects speak more with people. Spimes begin and end with data, so the stories they have to tell are comprehensive, ranging from their birth all the way through their death. Their memories will be far better than ours as humans, and essential to the future of the spime are the last two words of the above definition: "uniquely identifiable." For the spime to work, no two objects will have the same story. Even if something is mass-produced, it will no longer be "mass" in today's sense because it will be assigned an identity, just as RFID tags are already used to affix unique identities to billions of objects.

As Sterling points out, spimes do not exist yet, at least at a widespread level. But he identifies the adoption of RFIDs—or "arphids," as he calls the tags—as one of the first steps in building a spime future. As I cover throughout this book, examples already exist of pseudo-spimes that use RFID to at least partially communicate objects' stories, a capability also captured in the related concept of the "blogject" coined by computer scientist Julian Bleeker. The blogject is a portmanteau for "object that blogs" and communicates its state to infrastructure. But, as Bleeker points out, blogjects are about more than just tracking the mobility of an object or collecting sensor data about the environment. Blogjects can potentially rise to "first-class a-list producers of conversations in the same way that human bloggers do—by starting, maintaining and being critical attractors in conversations around topics that have relevance and meaning to others who have a stake in that discussion."[58] The conversations enabled through RFID sometimes rise to those levels, whether the conversations are about the temperature levels of food or the flows of cars on highway infrastructure.

Many examples of RFID covered throughout these chapters might not quite count as spimes or blogjects, but they represent precursors to a possible future in which the mobility of objects becomes comprehensive stories people can access. The stories then will likely affect decision making, and the stories will often focus on mobility and identification data. For instance, when someone uses RFID to reveal the authenticity of a pharmaceutical drug, they access a story that ensures an object remained unspoiled throughout its trajectory through the global supply chain. When someone uses RFID to take inventory and record each site a product passes through, they record a history of a specific objects' mobility. Later in this book, I return to the idea of spimes and blogjects to show how RFID enables objects to speak in new ways. In no small part, the ability to link objects to infrastructure, to give objects voices in an increasingly animated environment, has already been RFID's main contribution to the growth of the Internet of Things.

The Infrastructures of RFID

RFID's role in producing identification data shows how the technology is both a piece of infrastructure and a mobile technology. Infrastructure is a term used widely across multiple disciplines, with similar yet divergent meanings. Sociologists Susan Leigh Star and Karen Ruhleder posited that metaphors of infrastructure often present it as a substrate, or "something upon which something else 'runs' or 'operates.'"[59] They argue, however, that infrastructure is more than just that which supports other operations. Rather, "infrastructure is a fundamentally relational concept. It becomes infrastructure in relation to organized practices."[60] Simply building something does not automatically make it infrastructure. Rather, it becomes infrastructure through relationality that involves material (and/or discursive) structure as well as the practices it shapes. As Brian Larkin puts it, "As physical forms they shape the nature of a network, the speed and direction of its movement, its temporalities, and its vulnerability to breakdown."[61] Infrastructures are not just there; they are not neutral. They shape and enable behaviors while being shaped through their use. "Infrastructures are matter that enable the movement of other matter,"[62] and they do so in specific, political ways.

Infrastructures are obviously important, but they are not often flashy. Infrastructures are often designed to remain out of the way and fade into the background.[63] Some infrastructure, such as underground fiber-optic cable, is literally hidden.[64] Other forms of infrastructure, such as roads and cellular networks, become so mundane that people rarely pay attention to them until something goes wrong. Although the invisibility of infrastructure is sometimes overstated,[65] it remains true that many types of infrastructure affect people's lives in ways they pay little attention to. However, the mundanity of infrastructure masks its vital importance. The interpersonal computer-mediated communication analyzed by media studies scholars relies on large networks of infrastructure;[66] the mobility of people, ideas, and information is reliant on infrastructure; at a slightly more abstract level, economic analyses of trade patterns implicate the transnational infrastructures of the global economy. And equally important, infrastructures directly concern issues of power. Urban studies researchers Steve Graham and Simon Marvin's groundbreaking work *Splintering Urbanism* showed how "social biases have always been designed into urban infrastructure systems, whether intentionally or unintentionally."[67] Consequently, infrastructures often reflect and reify existing social relations, ranging from infrastructures of surveillance to infrastructures of mobility or, as I will show with RFID, infrastructures that meet at the convergence of the two.

Of course, one of the challenges of studying infrastructures is that they are often designed to not be noticed. Consequently,

Figure 1.1
An assortment of RFID tags. Photo courtesy of the author.

one of the goals of this book is to embrace an "infrastructural imagination" that encourages readers to delve into that which lies below the surface level of our surrounding space. I take that phrase from an interview between Brían Hanrahan and John Durham Peters in which Peters argues that "the infrastructuralist imagination ... seeks to appreciate all that is essential and off the radar."[68] I want to peel back the layers that hide RFID from the world. Peeling back those layers is particularly important for studying RFID because tags come in so many different shapes and sizes. Take the image in figure 1.1 as an example.

That image is a collection of RFID tags I collected in the course of my research. The tags range from small enough to be injected into a dog's back to larger tags with visibly printed antennae.

In chapter 3, I go into more detail on the capabilities of different types of RFID, but the important point here is that RFID tags come in many shapes and sizes depending on a variety of factors.

The tags themselves, even when one knows what to look for, can be difficult to identify because they are so varied. Adding a layer of complexity, the small size of RFID tags also lends themselves to some fairly intense forms of the invisibility of infrastructure. Tags can be injected into bodies (both human and animal), embedded in packaging, and sewn inside clothing. Many people reading this book are likely within reach of some kind of RFID embedded in between plastic, whether in the form of a contactless credit card or an access badge of some sort. Take the image in figure 1.2 as another example: a (now obviously broken) Washington, DC, SmarTrip metro card.

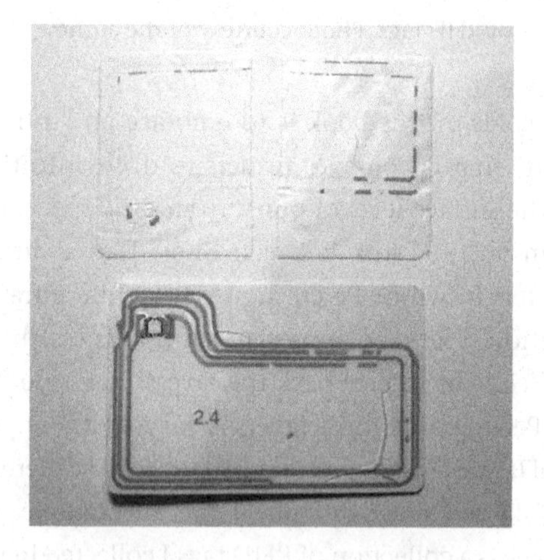

Figure 1.2
A disassembled DC Metro card that shows the internal RFID antenna. Photo courtesy of Tom Lee.

The card looks like any other card in someone's wallet, but one can peel back that surface layer to expose an RFID antenna that communicates with the card readers of the DC Metro system. Tearing off that outer plastic and examining the types of communication that occur beneath the surface is part of the infrastructural imagination that shapes this book. Those antennae embedded in plastic communicate; they link bodies' movements to infrastructure. They make each individual passenger uniquely identifiable. They are mobile communication.

The Structure of This Book

RFID technologies can be found all over the place, from the ears of cattle to the bodies of humans, from the plastic of subway cards to the labels of wine. These tags, regardless of where they are found or how they work, can make objects machine-readable, communicative, and uniquely identifiable. The tags produce data about objects that feed into larger systems for identification and analysis. But RFID is far from the only infrastructure of object identification. Consequently, chapter 2 looks at parallel technologies and links them to the role RFID plays in animating the physical environment. The chapter begins by discussing why barcodes are the most important antecedent technology to RFID and then looks at a related technology of identification that will play a crucial role in the Internet of Things: the IPv6 standard. After discussing related technologies, I then trace the history of RFID back to developments in radar and discuss the birth of the Internet of Things concept and some of the successes and failures of RFID adoption.

Chapter 3 goes into more detail on how RFID works. The chapter is not meant as a technical guide to the technology, and

I do not go into complicated explanations on the physics of radio waves or the intricacies of antennae design. Rather, I cover topics such as the difference between active and passive tags, RFID frequencies, and types of RFID data to push back against some of the misunderstandings about the technology. I argue that some articles confuse the technical capabilities of different types of RFID in ways that muddle the benefits and drawbacks of RFID adoption. The chapter also explains near-field communication—an emerging type of RFID that links smartphones to material infrastructure through radio waves. While chapter 3 goes into detail about how various types of RFID work, I do so in a way that is hopefully not too dry and will help readers be critical of claims made about what the technology can and cannot do.

Chapters 2 and 3 build a knowledge base about RFID's history and technical capabilities. Chapter 4 uses that base to delve more deeply into the role the technology plays in the Internet of Things, covering different conceptualizations of the Internet of Things and arguing for broad definitions that enable a more holistic understanding of the concept's potential. Additionally, a definition of the Internet of Things that includes "things" that do not have their own internet connection enables a deeper analysis of how object communication affects issues of space and place. Consequently, the chapter covers theories of the merging of physical space and computing before discussing multiple examples of the sometimes mundane, sometimes not, ways RFID is used to enable object communication.

Much of the excitement of the Internet of Things revolves around issues of data. By connecting objects and making them communicate, people will be able to collect data on previously impossible scales. And, going back to the original concept of RFID and the Internet of Things, object communication has always been about data. Attaching an RFID tag to everything from a book to a car turns an object into a machine-readable producer of data. Chapter 5 looks at RFID's role in producing data and how the technology both enables and is shaped by discourses about the supposed "big data revolution." The chapter examines the meanings of the word data and the epistemological shifts that accompany some embraces of big data. I then move on to examples of how RFID is used to produce data in diverse ways. The examples, however, do more than just look at different types of RFID data. Rather, the examples show the many factors external to specific technologies—ranging from types of software to regulations to religious texts—that shape what kinds of data are collected, tempering more utopian claims about data "speaking for itself."

Much of the data shared through RFID is about things. Billions of objects in the supply chain broadcast their identities; temperatures of cold chain transportation are transmitted through RFID and sensors. Other forms of RFID data, however, more significantly implicate issues of privacy and surveillance. As chapter 6 discusses, RFID became a target of privacy advocates in the mid-2000s, and some of the outcry about the implications of the technology likely slowed RFID adoption. The chapter looks at the privacy concerns of RFID in everything from tags on pairs of pants to tags on windshields to the RFID chip inside most passports. Importantly, chapter 6 uses the discussion from chapter 3 (of the affordances of different types of RFID) to push back against some alarmist accounts that overstate the technology's communicative capabilities. The chapter also urges the importance of an infrastructural imagination in understanding the oft-unnoticed types of object communication that play a role in shaping how identification infrastructures function.

The book concludes with a shorter chapter that looks into the future of RFID's role in the Internet of Things. I avoid making bold predictions about where RFID technologies will be in ten years, in no small part to avoid the embarrassment of being very wrong. But I do put recent developments—including Apple's removal of most near-field communication restrictions, the long-awaited "tipping point" in item-level tagging, and the automation of data collection—in perspective to show why this is an interesting time for infrastructures of identification. For the Internet of Things to reach its predicted potential, the types of object communication covered throughout this book will need to grow. RFID—in no small part because of its ubiquity, cheapness, and promiscuity—will be an enabling technology that gives a voice to billions of objects moving through the world.

Conclusion

The Internet of Things involves connecting objects and enabling them to communicate in new ways. The concept was first coined in a 1999 presentation about RFID technology, and the concept has matured since then. However, as useful as the concept of the Internet of Things may be at times, it can also hide more than it reveals. Namely, as I discuss in more detail in later chapters, the Internet of Things is almost always framed within a discourse of newness, as a technological development just off in the future and about to happen. That focus on the future can blind people to the many ways billions of objects already communicate directly with identification infrastructure. Transportation infrastructure connects with objects and responds when it senses their presence. In other words, our environment is already filled with connected objects and computing power. The Internet of

Things in some ways may still be emerging, but in other ways it is already here.

Consequently, the examples of RFID systems I cover throughout this book examine both the mundane and the experimental. I analyze transportation and access infrastructures that have been around since the 1990s. But I also look at RFID systems that are still in the pilot stage and might never achieve commercial viability. Ultimately a key part of what makes RFID such an interesting technology is its ubiquity and promiscuity. Once people know where to look, they can find tags all over the place. As with any relatively new technology, the analysis in this book can provide only a snapshot of a shifting landscape. Some of the topics I discuss will change as the technology matures or is replaced. The general focus of this book on identification and communication, however, will remain vital regardless of what happens in the coming years. For networks of object communication to grow, objects will have to be differentiated in the physical world. RFID is a key piece of how that differentiation happens, a piece that plays an often unacknowledged role in the functioning of everyday life.

2 Infrastructures of Identification

Researching RFID technology drew me into a world of mundane things. I read more than I could ever have imagined about how subway turnstiles work, how automatic tolls identify cars, how Google Pay works on phones, and how the tiny microchips in pets' backs identify them if they go to an animal shelter. The more I read, the more people I interviewed, the more layers I peeled back, the more I realized I was looking at the many ways the built environment differentiates things. The tollbooth differentiates one car from the millions of other cars; the RFID microchip in a dog's back differentiates one animal from another.

I then began thinking more about related technologies that work as infrastructures of identification. RFID, after all, is only one way objects are differentiated in the physical world, and it has its own history that is important to acknowledge. That history, however, cannot be confined merely to discussions of RFID technologies. First, RFID itself arose out of developments in radio and radar. Second, the development of RFID cannot be fully separated from similar technologies that work as infrastructures of identification. Consequently, this chapter provides background on two related technologies—barcodes (including QR codes) and the IPv6 standard—to better conceptualize how RFID works as an identification technology. After discussing barcodes and IPv6, I then move on to a relatively brief history of RFID that traces its roots back to radar and looks in some detail at the challenges and successes of the RFID industry.

The first technology of identification covered in this chapter is barcodes, which are likely the most important antecedent to RFID technology and a prominent automatic identification and data capture (AIDC) technology. AIDC "refers to the methods of automatically identifying objects, collecting data about them, and entering that data directly into computer systems,"[1] and AIDC techniques include RFID, biometrics, and other methods. Barcodes are still the most widely used AIDC technology and have become so ubiquitous that they just seem to be a given part of the physical world. But barcodes have their own history and had major influence on inventory and logistics, and they provide a process of object identification that RFID builds on in various ways. Consequently, barcodes are an important mobile technology that provides insight into both the improved technical capabilities of RFID and the roles standards play in creating and stabilizing identification infrastructure.

Barcodes are ubiquitous and often ignored, but at least they are visible. Another identification technology related to RFID is even more invisible than antennae embedded between plastic: IP addresses. RFID is an animating technology of the Internet of Things because of its ability to make billions of objects uniquely identifiable. But RFID is only one part of the Internet of Things. As I discuss in more detail in chapter 4, many "things" will have their own internet connection and will not use RFID as a communicative mediator. Each of these devices will need to be uniquely identifiable and will be assigned its own IP address. The current dominant internet standard—Internet Protocol version

4 (IPv4)—does not have the naming capacity to connect billions of additional objects to the internet. Consequently, the main contribution of IPv6 is expanded identification capacity, making IPv6 a parallel infrastructure to the increased data capacity of RFID in the Internet of Things.

Part of what makes barcodes, IP addresses, and a history of RFID interesting is their mundanity and the often hidden role they play in everything from inventory to interpersonal communication. Barcodes and IP addresses are major parts of the back-end that make our world work. Just today, the groceries I bought, the package I had delivered, the information I looked up, and the messages I sent my significant other were all enabled by the identification infrastructures covered in this chapter. While the next chapter goes into more detail on how RFID technology works, this chapter provides necessary background for getting at how the concept of object communication extends past RFID to related infrastructures of identification.

Barcodes and the Universal Product Code

Barcodes are one of the most transformative mobile technologies few people ever talk about. As Nigel Thrift argued, "The barcode is a crucial element in the new way of the world, one which remains largely untold."[2] Someone my age (mid-thirties) who has grown up in the United States likely has no memories of a time before ubiquitous barcodes. By the time I have memories of being in a grocery store, almost all items appeared to have barcodes, and prices were calculated through optical scanners that read the object. Before I started researching RFID, I had never stopped to think how retail worked before the barcode. The technology was just a given part of the world, not a part with a unique history and important agential effects on inventory and supply. As sociologist Nancy Baym has argued, dominant technologies eventually "become so taken for granted they are all but invisible."[3] Barcodes are a prime example of that tendency.

The goal of this discussion is to push back on that pseudo-invisibility by looking in some detail at the barcode as a mobile technology. Barcodes are, after all, the most important technological antecedent to contemporary RFID technologies, though they work differently and rely on optical scanning rather than radio waves. To understand some of the social impacts of RFID technologies, it is first necessary to reflect on the history of barcodes and the Universal Product Code (UPC) and International Article Number (EAN).

In 1948, Bernard Silver, a graduate student at Drexel University, overheard a supermarket manager ask a Drexel faculty member about technologies that could improve retail checkout. Silver was intrigued and contacted his friend Joseph Woodland, and the two began working on a solution.[4] They came up with one that was inspired by the dots and dashes of Morse code but with different-shaped lines rather than pulses. The two men then filed a patent in 1949 for a close relative to the modern barcode (see figure 2.1). However, the skeleton of the barcode was ahead of its time because it was invented about eleven years before its key companion technology: the laser.

The first laser was built in 1960 by Theodore H. Maiman, and lasers made it possible to optically read the lines on barcodes. By the mid-1960s, the railroad industry had begun to adopt a relative to the barcode called KarTrak, and by 1967 the industry had created a national coding standard. Then, in 1969, a company called Computer Identics deployed what was possibly the "first true barcode system anywhere."[5] The system was deployed

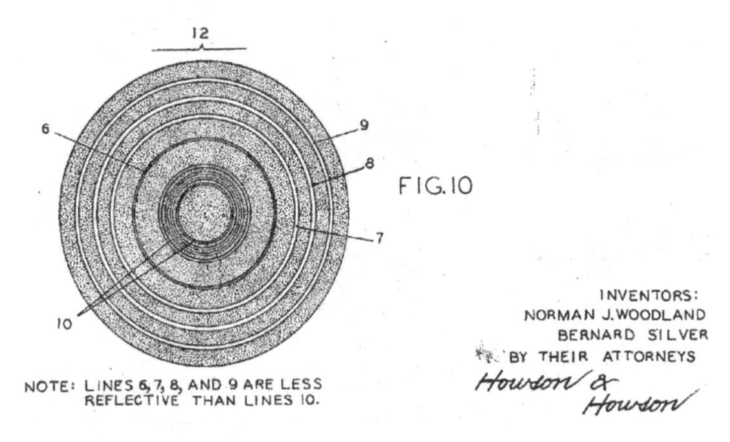

Figure 2.1
The bull's-eye-shaped barcode design in Woodland and Silver's original patent. Source: US Patent 2, 612, 994.

in a General Motors plant in Michigan and a distribution plant in Nevada.

The breakthrough that reshaped much of the world's economy was yet to come. The eventual triumph of the barcode had just as much to do with standardization and human cooperation as it did with technical features. The cooperation necessary for standardization came from an ad hoc committee that first met in August 1970 and consisted mostly of representatives of the Grocery Manufacturers of America and the National Industry of Food Chains.[6] The goal of the committee was to determine whether "a standard industry product code" was worthwhile and, "if so, what should that code be?"[7]

Once the committee determined that a standard code would benefit the US grocery industry, it then had to choose the technology that should distribute the code. Here is where the barcode came in. There was little debate about using barcodes. There was

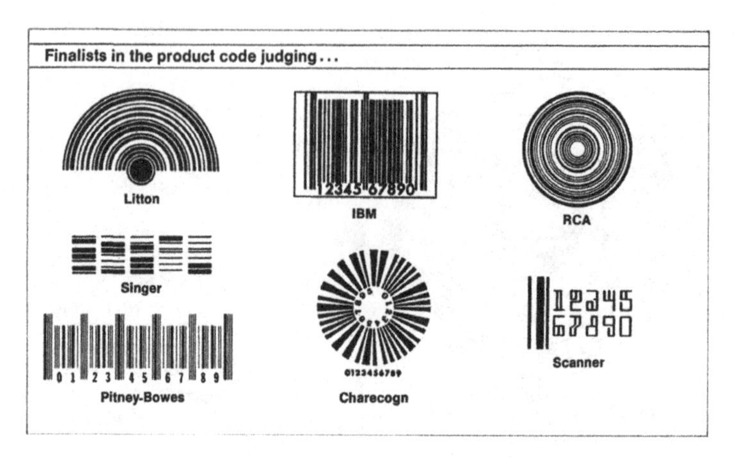

Figure 2.2
The barcode finalists. Image reprinted with permission from the IDHIstory .com museum.

debate, however, about how many digits the barcode should carry and what the barcode should look like. To pick a design, the committee formed a symbol selection committee that evaluated different proposals (see figure 2.2). Two of the major proposals came from RCA and IBM. RCA proposed a design that would be unrecognizable to most people today, focusing on a bull's-eye-shaped code. IBM's proposal focused on the vertical-line barcode technology still commonly used. Near the end of the selection process, the committee "achieved a consensus that the fundamental choice was between the bull's eye symbol proposed by RCA and Litton, and the oversquare proposal of IBM."[8] In March 1973, the committee voted and chose IBM's proposal over RCA's bull's-eye.

Once the committee had chosen the barcode design, it had to settle on a numbering standard. After a few years of effort, the Universal Product Code (UPC) was born. The UPC is a specific type of barcode symbology still widely used around the world. A UPC includes twelve digits, and the set of numbers corresponds to information that was determined and maintained by the Uniform Code Council (UCC), a council developed out of the ad hoc committee and now known as GS1. A sample UPC is a series of numbers:

0 23746 **89343** <u>8</u>

In the code above, the first set of numbers (italicized) represents the manufacturer code, a code assigned by the GS1 (previously the UCC) that shows which organization the product comes from (e.g., Kellogg's). The second set of numbers (the bold numbers) is the item number (e.g., Corn Flakes). The manufacturer is responsible for making sure item numbers are assigned only to individual products. The final digit (underlined) is the "check digit" that makes sure the product was scanned correctly.[9]

So here's how the UPC works: An optical scanner scans the lines of the barcode and detects the unique pattern of distinct lines and the organization of white spaces. The UPC is coded within that pattern, and the database attached to the scanner matches the manufacturer and the item. For example, a single UPC might identify an item as Kellogg's Corn Flakes. And here is the most important point specifically related to RFID: UPC barcodes identify *classes* of products, not individual products. The Kellogg's Corn Flakes are identified, but as far as the UPC and the database is concerned, each box of Corn Flakes is the same as every other box of Corn Flakes. Barcodes do not have the storage capacity to uniquely identify every single object. In addition, UPC (and EAN, as I discuss later) barcodes only identify the object. They do not contain additional information such as price or sell-by date; those data points are stored in the database that received the identifying information. As Martin Dodge and Rob Kitchin described, "The strengths of the barcode (simplicity and lack of granularity) are also its major weakness and, as a result, barcodes are being replaced by new coding systems held by smart labels and tags."[10]

The UPC is still one of the dominant barcode forms, but it is not the only barcode standard that reshaped logistics and supply in various industries. The UPC was designed specifically for the US grocery industry. A consortium in Europe created the International Article Number (also known as the European Article Number, EAN), which is similar to the UPC but contains thirteen digits.[11] At first, the difference between the two standards led to scanning problems at the point of sale, but "by the early 1980s all scanning equipment marketed in the US could scan either UPC or EAN symbols."[12] Japan also created a Japanese Article Number (JAN) in 1978 that is compatible with the EAN. As a final note, many barcodes do not use the UPC, EAN, or JAN, and other standards—such as the Codabar Monarch or the code 128—enable other numbering schemes.[13] In addition, some organizations have developed their own numbering standards. For example, FedEx and UPS have proprietary barcode systems that refer back to uniquely ordered identification systems.

With relatively consistent standardization, barcodes spread fairly quickly. In 1974, the first UPC was scanned on a pack of gum in a grocery store in Troy, Michigan. Within a few years, 75 percent of products in supermarkets in the United States carried barcodes.[14] Not long after, barcodes spread into retail in Europe and the United States and began to be used outside point-of-sale interactions as a logistical tool of identification. While there is no definitive way to know just how many barcodes are used today, the GS1 database now has more than one million registered companies, and barcodes are scanned over five billion times a day.[15]

The Social Consequences of Barcodes

As a key infrastructural technology, barcodes are a ubiquitous, mostly ignored part of the physical world that exerts a major influence on processes of identification. Most obviously, barcodes make objects machine-readable. They are "bit structures" that are "the mechanism by which the virtual establishes its logic in the real."[16] For all the discussion of how smartphones and locative media have led to hybrid spaces in which the digital and physical merge in new ways,[17] barcodes were an unacknowledged but crucial forerunner to the hybridity of our contemporary world. They made physical objects digitally identifiable, bridging the gap between the materiality of objects and the digitality of back-end databases.

The debates are mostly lost to time, but the bridging of the physical and digital through barcodes was controversial when the technology was first introduced. As Stephen Brown describes in his history of the UPC, the members of the ad hoc committee did not anticipate backlash to barcode technology,[18] and they were surprised when that backlash came from consumer groups. To a significant degree, the backlash concerned issues of technological agency: consumer groups worried that people were ceding control of pricing to machines. Instead of watching a clerk type in the price of a product, the product was scanned and the price was automatically pulled from the database.

The height of the early backlash came in 1974, when the UPC was released. Carol Tucker Foreman, then president of the Consumer Federation of America, went on the *Phil Donahue Show* and criticized the UPC because of its lack of financial transparency.[19]

The push against the UPC picked up from there, and consumer groups allied themselves with labor unions concerned with the loss of jobs because of barcode scanning. The protests resulted in a compromise between consumer groups and the grocery industry: grocery stores agreed to keep price tags on most products rather than replacing them fully with a barcode. The barcode would still replace the process of a human manually typing in a price, but consumers would be able to see the price rather than have it tied solely to the link between the UPC and the database.

Eventually much of the backlash died down as the barcode became more accepted and more ubiquitous. People became used to ceding human agency to the little black lines found on their products. From there, barcodes had major impacts on the grocery and retail industries, similar in some ways to how proponents hope RFID will reshape logistics and retail. Namely, barcodes (and their standardization) enabled companies to "move goods through a retail network with unprecedented speed, accuracy, and efficiency, ... learn from buying behavior, ... and respond to actual demand in a manner that would have been impossible in earlier times."[20] Barcodes not only made physical objects digitally identifiable, they also made them trackable and recordable. Each object read by a scanner left a digital trace that, if properly managed, enabled manufacturers and sellers to use the "bit structures" produced through barcodes to drive decisions.

Barcodes are likely the key infrastructure of identification of the second half of the twentieth century. Buried in their black lines, barcodes communicate numbers to classify and sort objects on store floors and in supply chains. Crucially, their identification capabilities enabled new practices. Because items could be differentiated and sales more easily logged, barcodes subtly shifted relationships between objects and manufacturers and even between human bodies and businesses. Objects on store floors became database entries that could be logged through optical scanners. People's purchasing behaviors could be tracked by compiling individual barcode reads and attaching them in a database to the barcodes found on supermarket loyalty cards, which themselves have been a target of surveillance critics.[21] A supermarket loyalty card is a piece of identification attached to an individual shopper, and the totality of barcode reads then become entries in that person's digital profile compiled by the loyalty program. Barcodes are an example of how infrastructures of identification turn objects into data and then use that data to reshape and refine existing practices.

One way of thinking through the social impacts of barcodes is through communication scholar James Beniger's concept of the "control revolution."[22] Beniger's work traced the development of various techniques, infrastructures, and media technologies in the nineteenth and twentieth centuries. His unique contribution was recognizing that all these areas—whether the ledgers that tracked freight cars or the birth of modern computing—embodied techniques to exert control. The history he tells is one of repeated control crises in which new techniques and technologies had to be developed to record, process, and analyze growing amounts of data. Beniger's discussion did not include barcodes, but it easily could have. Barcodes helped companies insert new forms of control into networks of retail. The technology made products trackable and traceable, which had real effects within a few years of the release of the UPC and the EAN. Two consequences of control were particularly notable: a "striking pattern of increased concentration" in the grocery sector, and a "remarkable proliferation of products."[23] Companies were able to control and improve efficiency, leading to higher profits, which were fed back into development and helped contribute to increased concentration. The technology also lowered inventory requirements, made more granular product tracking possible, and "automated the record-keeping tasks associated with variety."[24]

In essence, recognizing two of the major capabilities of barcodes is essential to understanding the later analysis of RFID: (1) they are a bridge between the physical and digital that makes objects machine-readable and partially identifiable, and (2) they produce data that feeds into systems that enable increased control over inventory and point-of-sale interactions. RFID would later expand on the communicative potential of barcodes, though barcodes still remain the dominant AIDC technology. As the discussions that follow explain, barcodes also expanded beyond retail and supply to bridge gaps between mobile phones and the physical world, showcasing a more democratized form of machine-readability that is an important precursor to the near-field communication (NFC) technology I discuss in chapter 3.

QR Codes

QR codes are a type of two-dimensional barcode. Traditional barcodes are one-dimensional, meaning they contain their information horizontally. Two-dimensional barcodes—including but not limited to QR codes—contain information both vertically and horizontally, most often in a square form rather than the rectangular one-dimensional form of a barcode. Consequently, QR codes can contain more data than the twelve-digit UPC or the thirteen-digit EAN.

Here it becomes important to clear up some terminology. The phrase "QR Code" is the trademarked name of one of many matrix-type 2-D barcodes. The trademarked name refers to a specific type of 2-D barcode first designed to be used in the Japanese automotive industry. The QR code then spread outside the automotive industry and, in a few cases, though fewer than proponents had hoped,[25] replaced traditional barcodes because QR codes had faster readability and could contain more information. If QR codes had stayed exclusively in the automotive industry, they would not have gotten much attention in humanistic and social scientific research; they are just a more advanced barcode that carries more data. However, when most people use the phrase "QR codes," they do not mean a trademarked barcode built for the automotive industry. Instead, as media scholar Leopoldina Fortunati argued, QR codes "are meant to be read by smartphones through specialized applications."[26] In other words, the term QR code became a stand-in for 2-D barcodes people could scan with their smartphone to follow the data—most often a URL—embedded in the code.

QR codes work as a more data-rich barcode and, "with the mediation of the mobile phone, can then act as a bridge between a static media (like a print newspaper) and a dynamic media."[27] Newspapers include QR codes people can access to go to videos about stories, printed flyers include QR codes that take people to mobile applications they can download, museums include QR codes people can scan to learn more about an exhibit, and marketers include QR codes that allow people to access digital coupons. As a study of QR code adoption in Japan found, "The primary benefit of QR code is its pivotal role as a bridge between offline and mobile media."[28]

For all the attention QR codes received in the mid-2000s, they have been somewhat of a disappointment.[29] One of the problems QR codes have faced is a rather straightforward issue of usability. Downloading an application and holding a phone up to a flyer or newspaper requires significant buy-in. QR codes can be scanned quickly, but they are often not the most convenient way of delivering information. In addition, the usage of QR codes is too often "rooted in a simple vision: the more communication there is, the better it is."[30] The codes have the potential to link print to more information-rich digital environments, but they are often used in situations in which people do not necessarily want a more information-rich environment—they become obligatory passage points for an end destination too few people want to go. Take newspapers, for example. Research shows that most people skim print papers for news and devote little time to in-depth reading. Consequently, people already have too much information to process; adding QR codes to an article asks for even greater investment of time and effort. The same is true of posters, flyers, or advertisements that include QR codes. Only people already invested will take the time to find even more information when confronted by the increasingly dense layers of information in our physical environment.

Of course, just because QR codes have not yet been as widely adopted as some hoped does not mean they never will be. These codes have agency in networks that connect print to digital; they shape how that process occurs and, if removed from the network, the connection disappears. But another problem QR codes face is more relevant to the later chapters of this book: namely, they have begun to be replaced in some cases by NFC technology that enables smartphones to read tags that can be connected to print objects.[31] Just as QR codes moved barcodes from business-oriented uses to end users, NFC has the potential to alter how people interact with RFID technologies.

IPv6 as an Infrastructure of Identification

Most of the internet is still routed using the Internet Protocol version 4 (IPv4).[32] IPv4 was first described in a 1981 report and, more than thirty years later, is still the dominant internet protocol.[33] IPv4 regulates IP addresses, which are the identification assigned to devices connected to the internet. When a computer connects to the internet through an internet service provider (ISP), the ISP assigns an IP address that makes the device uniquely identifiable and works similar to a mailing address. IP addresses provide identification capabilities that make it possible to trace information requests to individual computers.

The problem with IPv4 is that, kind of like barcodes, it has limited identification capacity. IPv4 addresses use a 32-bit identification system, which allows for roughly 4.3 billion unique IP addresses.[34] In the early 1980s, when IPv4 was developed, few devices were connected to the internet, so that limitation did not matter; however, with the rapid increase in internet-enabled devices, IPv4 addresses have essentially been depleted.[35] In the Internet of Things, more and more devices will be connected to the internet, further overtaxing the IPv4.

People recognized the potential exhaustion of IPv4 addresses by the early 1990s, and the Internet Engineering Task Force (IETF) announced a call for white papers about a new standard in 1992.[36] Then, in 1995, the IETF began releasing a series of request for comments (RFCs), beginning with RFC 1883, which defined IPv6 as "a new version of the Internet Protocol, designed as a successor to IP version 4."[37] The major change from IPv4 to IPv6, and the first listed difference in RFC 1883, is IPv6's "expanded addressing capabilities." IPv6 expands the 32-bit IP addresses of IPv4 to 128 bits, which provides 2^{128} naming options, or, to put it in terms that are almost too large to comprehend, 340,282,366,920, 938,463,463,374,607,431,768,211,456 unique IP addresses.

IPv6 has other differences from IPv4, including changes to heading formats and improved privacy capabilities. The expanded number of addresses is the major difference, however, and an important part of the growth of the Internet of Things. If experts are correct in predicting that over twenty billion devices will be connected to the internet by 2020,[38] those devices will need IP addresses, and IPv4 cannot handle that kind of capacity. IPv6, on the other hand, provides more than enough naming capability for the internet to expand into everyday objects. Consequently, the IPv6 standard is about expanded identification, just as RFID extends the identification capabilities of barcodes.

IPv6 and RFID have other similarities as well. The IPv6 standard is in many ways superior to IPv4, just as RFID enables more affordances than barcodes. But, as I examine later with RFID, the rollout of IPv6 has been bumpy. IPv6 has been available for two decades, and agencies have already begun to face problems with the limited naming capacity of IPv4. As of 2016, however, IPv6 had achieved only 10 percent deployment, which is better than it sounds, considering that the number was at 6 percent only a year before.[39] So why has IPv6 been so slow to be implemented? Interestingly, a few of the major reasons apply also to RFID, showing why supposedly "more advanced" technologies do not automatically win out. First, switching to IPv6 is expensive, and most internet infrastructure (e.g., routers) was built for IPv4 (just like most companies already have established barcode infrastructure). Second, just as companies cannot use barcode scanners to read RFID, the IETF did not make IPv6 backward-compatible with IPv4.[40] Consequently, systems need to use dual stacking to be compatible with both IPv4 and IPv6 addresses. The massive amount of work that goes into switching from IPv4 to IPv6 means the two will have to coexist for at least the next few years, just as some systems now blend barcodes and RFID.

Despite the slowness of the rollout, IPv6 will eventually overtake IPv4, in no small part because it is difficult to imagine an Internet of Things without IPv6. While this book is about RFID's role in the Internet of Things, IPv6 is relevant because, despite obvious differences, it addresses some of the same issues as RFID deployments: namely, IPv6 can be conceptualized as an infrastructure of identification, albeit an altered type of identification. IPv4 does not have the capabilities to assign unique IP addresses to tens of billions of objects, just as barcodes do not have the storage capacity to assign unique UPCs or EANs to tens of billions of objects.

To be clear, IPv6 and RFID are parallel technologies of identification and are not reliant on each other. IPv6 would still be necessary if RFID went away, and likewise, RFID's development is not directly related to the growth of IPv6. After all, the growth of barcodes had little to do with the popularization of IP addresses. However, while they form different types of infrastructures of identification, RFID and IPv6 both showcase one of the major shifts tracked throughout this book: namely, vastly expanded processes of identification are a shaping force of the contemporary world. Building an animated environment of tens of billions of connected things will require the sorting and differentiation of objects on unprecedented levels, and RFID and IPv6 will each do their part as key infrastructure of identification.

A Brief History of RFID

One of the challenges of writing a fairly brief history of RFID is choosing where to begin. Obviously, RFID would not exist without the invention and development of radio in the late nineteenth century by people like Hertz, Popov, and Marconi. But I want to start at a different point: the development of radar.[41] In 1904, German physicist Christian Hülsmeyer performed the first experiment that used radio waves to detect objects. Hülsmeyer's invention was followed by a series of developments by scientists in Europe, Japan, and the United States using radio waves to detect and identify objects.[42] By 1934, scientists working in the US Naval Research Laboratory tested an experimental radar system that was used by the US Army the following year. Then, in 1935, a British Air Ministry team led by Robert Watson-Watt patented a similar radar prototype that eventually developed into the radar system that helped Britain win the Battle of Britain.[43] By 1940, radar was in wide use in World War II and is still used today.

So what does radar have to do with RFID? Radar works by sending out radio waves that keep going until they hit something and then bounce back and are received by the radar system. The system then uses complex calculations to determine the size of the object and the speed at which it is traveling. In other words, radar uses radio waves to detect and identify objects. The roots of RFID are found in the design of radar systems, particularly in a fairly unnoticed-at-the-time article about radar published by Harry Stockman in 1948.[44] Stockman was an expert in radar and taught classes on radar during World War II. After the war ended, he received his PhD and took a position as the head of the Cambridge Field Station's Communications Department.

While serving in that position, Stockman wrote the paper often credited as the "genesis" of RFID,[45] "Communications by Means of Reflected Power."[46] The paper described an experimental arrangement that

> created no less than a crude, pre-digital form of passive backscatter modulation. By modulating numerals onto a sensitive radar beam, [Stockman] devised a system of automatically identifying objects by radio waves, and along with that, he turned seeing into reading. Radar became reader.[47]

The experiment described by Stockman used a "number identification target system" that could uniquely identify objects based on revolution speed. His system was still far from the transponder systems used in contemporary RFID systems, but it foregrounded the growth of backscatter communication, which is a method of communication used by passive RFID tags that do not have their own batteries. In Stockman's abstract, for example, he wrote about "carrier power generated at the receiving end and the transmitter replaced by a modulated reflector."[48] The reflector in this case is seen by some as a precursor to contemporary passive RFID tags (described in more detail in chapter 3). The reflector has no power source in itself and instead draws from the signal of the receiving end.

Stockman's description of a backscatter system might represent the "first work exploring RFID,"[49] but it took almost thirty years for all the technical pieces—microprocessors, transistors, and integrated circuits—to be developed to enable the start of commercial RFID development. A few major conceptual works in the 1960s by people like R. F. Harrington and Robert Richardson developed a knowledge base on which later RFID systems would be built, but it was not until the 1970s that developers, inventors, companies, academic institutions, and government laboratories were actively working on RFID, and notable advances were being realized at research laboratories and academic institutions such as Los Alamos Scientific Laboratory, Northwestern University, and the Microwave Institute Foundation in Sweden.[50]

The 1970s also saw an increase in corporate interest in the technology, with companies like RCA and Raytheon developing rudimentary systems. The Port Authority of New York also began testing systems, and RFID was developed (though mostly in experimental terms) in areas such as animal tracking, transportation, and automation. Then the 1980s saw a few commercial deployments of RFID in sectors in the United States and Europe. Importantly, the growth of RFID in the 1980s was not just because of improvements in radio technologies: the growth was also linked to cheaper computing that made it possible to handle large amounts of RFID data.

Although the 1980s saw some of the first operational RFID systems, deployment was mostly limited to transportation in the form of automated toll collection. In 1987, Norway released the first commercial RFID toll-collection system, which was followed two years later by the first United States RFID toll system in Dallas, Texas.[51] The 1990s then saw the stabilization of RFID technologies in certain industries.[52] For example,

- The first open toll system that enabled people to pass through at highway speeds was released in Oklahoma in 1991 and was followed by the rapid dispersion of RFID toll collection in North America, Europe, and parts of Asia.[53]
- Rail companies began tagging cars with RFID.
- Companies began developing mature RFID applications in access control and animal tagging.[54]

- RFID began to be used in "smart" public transportation systems that used RFID-enabled cards to enable access and monitor ridership. The first system to rely on RFID was the South Korean Upass system (introduced in 1996), followed closely by Hong Kong's Octopus card (introduced in 1997), which was the inspiration for the London Oyster card.[55]
- Companies began to develop contactless payment systems. In 1997, for example, Mobil released the Speedpass, which people could use to pay for gas.[56] The Speedpass, which still exists to this day, was an early example of a contactless payment card.
- In the mid-1990s, IBM filed the first patent for an ultrahigh frequency (UHF) RFID system, which was a precursor to the UHF systems I examine in chapter 3.[57] UHF RFID is the technology most frequently associated with the item-level tagging covered in the discussion that follows.

The 1990s was the decade that RFID began to mature as identification infrastructure in payment systems, animal tagging, access control, and transportation. Still yet to come, however, was one of the major developments that would shape the RFID industry.

1999: The Internet of Things

The same year the first book devoted solely to RFID technology was published—by Klaus Finkenzeller in 1999[58]—an idea that would change the RFID industry was formed: the idea of the Internet of Things. For all the talk now of individual IP addresses, the original concept of the Internet of Things was based on RFID technology. Kevin Ashton coined the term in a presentation to Procter & Gamble executives that detailed how to use RFID to individually identify billions of individual items (see figure 2.3).[59] Ashton imagined a future somewhat like

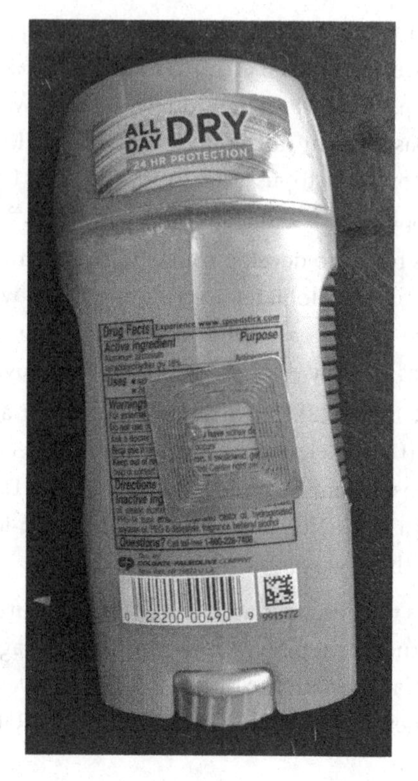

Figure 2.3
A random RFID tag I found on a Speed Stick deodorant in the grocery store. Photo courtesy of the author.

Sterling and his spimes, in which individual items would be able to "speak" to infrastructure connected to back-end databases. His argument was that people needed to move away from human-focused data to include "things" in the conversation, and the best way to do that was through RFID. The linking of objects to infrastructure through passive RFID is an idea that still shapes the Internet of Things to this day.

In addition to the creation of the initial Internet of Things concept, 1999 was also the year MIT's Auto-ID Center was formed. The Auto-ID Center focused on the development of RFID technology and was created by Ashton and three MIT professors, Sanjay Sharma, David Brock, and Sunny Siu. The center was funded by the Uniform Code Council (UCC), Procter & Gamble, and Gillette. The UCC's involvement in the center was particularly notable because 1999 was the twenty-fifth anniversary of the scanning of the first UPC barcode. When the symbol-selection committee settled on the UPC, they expected the barcode to last twenty-five years. Not coincidentally, the Auto-ID Center was "launched on September 30, 1999, during the twenty-fifth anniversary celebration for the bar code at the Smithsonian Museum in Washington, D.C."[60]

The early years of MIT's Auto-ID Center focused on two main ideas: changing the way people thought about RFID and developing the Electronic Product Code (EPC) system. Throughout much of their relatively short history, RFID tags had been viewed as "a mobile database that carried information about the product or container they were on with them as they traveled."[61] They were programmed to carry detailed data that had to be parsed, which often led to significantly higher-priced chips. Sharma and Brock focused on low-cost RFID tags that contained only a serial number, with the data about the tag stored elsewhere, much like the way the UPC on the barcode links back to a database entry and does not contain much information in the code. Essentially, "Sharma and Brock turned RFID into a networking technology by linking objects to the Internet through the tag."[62] Rather than treat an RFID tag as a mini database of a product's history, the tag would instead be the link to bridge the gap between the physical material and the object's digital identity, showcasing an early example of the Internet of Things concept I examine in more detail in chapter 4.

Another major accomplishment of the Auto-ID Center was the establishment of the EPC system: "A suite of standards and technologies that weaves basic RFID into a standardized scheme for keeping track of material in the supply chain."[63] While the full EPC system includes various technical standards—most importantly, how tags and readers communicate with one another—the numbering part of the EPC is RFID's version of the UPC or EAN. Just like with barcodes, many RFID systems do not use the EPC, but the standard was designed to be used mainly in product tagging. Just as the UPC and EAN were crucial to barcode adoption, the EPC helped stabilize RFID in retail and the supply chain. Consequently, the next section goes into detail on how the EPC works and how it compares to earlier barcode standards.

The Electronic Product Code

Almost by definition, standards are dry and designed to not be noticed. Standards are nonetheless a crucial, often ignored, part of our everyday world. The internet would not work without a host of standards that remain mostly invisible to end users. My house would not be built the way it is without standards. My food would not be as safe without standards. Barcodes, as discussed above, needed the standardization of UPCs and EANs to take off and work across industries. In effect, "standards are where language and world meet" and are "about the ways in which we order ourselves, other people, things, processes, numbers, and even language itself."[64]

Barcodes can contain only a small amount of data compared with RFID; consequently, RFID technology required a new numbering system that took advantage of its technological affordances. The UPC and EAN include twelve and thirteen numbers, respectively, with different sets of numbers identifying different parts of a product. The EPC is based on a similar structure but contains far more data. Therefore, as defined by the organization EPCglobal,[65] whereas a UPC identifies classes of objects, an EPC is "a universal identifier for any physical object. It is used in information systems that need to track or otherwise refer to physical objects."[66] The key part of that definition is found in the first clause: "identifier for *any* physical object."[67] While there are different uses for the EPC and more complicated explanations of how the code works as an identifier, what I want to focus on here is how the 96-bit code works to extend the identification capabilities of barcodes.[68] The number shown here is a sample EPC:

01. *0000A82* . 00014D . ***00195EOF***

The bold text (01) represent the header, which determines the numbering structure. The italicized text (*0000A82*) is the domain manager, which typically identifies the product manufacturer. The underlined text (00014D) is the object class, used to identify groups of products. The final text (***00195EOF***) is possibly the most consequential piece of the increased data capacity: the serial number. The serial number enables the identification of unique objects within a larger object class. Table 2.1 breaks down just how many combinations are available using the EPC standard.[69]

Table 2.1

Domain manager (typically manufacturers)	267,435,455
Object class (types of objects)	16,777,215
Serial number (unique instantiations of the object)	68,719,476,735
Total Combinations	2,541,865,828,329

In sum, EPCs provide "a number range large enough to uniquely identify every object on the planet."[70] The sheer number of combinations is rather remarkable. For as little as around ten cents a tag, every single item in a supply chain can be differentiated, sorted, and tracked through identification data.

The development of the EPC is a crucial piece of recent RFID history. The language of standards represents a different type of infrastructure than readers built in the environment or even software used to filter RFID data, but they show how infrastructure can be layered and relational. The material infrastructure of RFID systems in the supply chain rely on the discursive infrastructure of standardization to grow; without a dominant standard, it can be difficult for an industry to expand because of uncertainty from the people who have to buy into the emerging technology. In a case study of ten major corporations that have adopted RFID, University of Leicester researcher Adrian Beck found that "all agreed that without them [EPC standards], it would be more difficult to innovate and evolve in the future."[71] In particular, the participants in the study pointed out that standardization kept companies from getting locked into certain providers, and standards reduced confusion when formatting tags. For example, imagine a supplier tagging items with RFID. The supplier ships items to three sites, but each site requires a different data format. RFID adoption would come to a halt because an already complex process would become impossibly complex without standardization. Take barcodes as a historical example: barcodes could not have spread as quickly if every grocery store demanded a different data standard.

As a final note on the EPC, just as barcodes do not have to contain a UPC or EAN, many forms of RFID covered in this book do not use an EPC. An RFID-enabled toll tag, for example, transmits a "unique serial number associated with a user's personal account."[72] That serial number often relies on a proprietary standard, which is why a toll tag for one system (e.g., the E-ZPass) will not work with a competing system (e.g., FasTrak). There are also many standards governing RFID usage—such as RFID smart cards, access badges, and RFID used in sensor networks—that are approved through the International Organization for Standardization (ISO). Regardless, while many RFID tags do not use EPC, the standard was necessary for the progression of RFID as a supply-chain technology, particularly with item-level tagging. In addition, the EPC shows just how important a role identification plays in the various forms of object communication covered in this book. The main goal of the EPC was to create a numbering standard that can be applied to an almost incomprehensible number of objects. Just as the IPv6 addressed the identification limitations of the IPv4, the EPC did the same with the already ubiquitous UPC and EAN.

The Contemporary State of RFID

By 2003 the EPC was established and the Auto-ID Center had gained support from more than a hundred large companies and the US Department of Defense. That success meant the center was now responsible for both the commercialization of the EPC and research about RFID, and in 2003 the center divided into two branches: EPCglobal and the Auto-ID Labs. EPCglobal is a GS1 initiative responsible for maintaining the Electronic Product Code and some standards governing RFID protocols, much like the UCC and EAN International were responsible for governing barcode technology.[73] The research part of the Auto-ID Center led to the creation of the Auto-ID Labs, which is a global network of seven labs that focus on RFID and the Internet of Things. The

Auto-ID Labs claim to be the "leading global research network of academic laboratories in the field of Internet of Things" and have continued the push to explore the use of RFID to interlink objects.[74]

With the formation of EPCglobal and the Auto-ID Labs, the future of RFID looked bright. In November 2002, Gillette placed an order for five hundred million EPC tags, showing that "EPC technology had arrived in the real world."[75] The future for RFID tagging in the supply chain then looked even brighter when Walmart—the largest company in the world by revenue and the largest grocery supplier in the United States—introduced a major RFID mandate in 2003. The mandate shifted somewhat throughout 2003 and eventually specified that all suppliers to Walmart had to tag pallets of goods (not individual items) with RFID by the end of 2006.[76] The top one hundred suppliers had to tag pallets starting in January 2005. The mandate was a huge deal in the burgeoning RFID industry. It showed that RFID tagging now had massive support from possibly the world's most powerful retailer and led to bold predictions about RFID having reached a "tipping point" on its way to mass adoption.[77] The Walmart mandate was then followed by an RFID mandate from the US Department of Defense and similar mandates from other retailers such as Target and Metro.[78]

The early and mid-2000s period of the RFID industry show why the oft-stated adage about the "invisibility of infrastructure" can sometimes be a bit overstated. During the period surrounding the Walmart mandate, RFID was an infrastructure written about in many popular sources and as I explore in chapters 4, 5, and 6 had already received the attention of conspiracy theorists, evangelical Christians, and privacy advocates. While the individual tags used to identify items might have remained fairly invisible, the larger narrative surrounding the technology was anything but.

In retrospect, however, the Walmart mandate likely raised expectations for RFID too quickly. Many manufacturers did not react positively to the mandate, arguing that implementing large-scale RFID systems was too costly and provided them with little benefit. In addition, passive RFID tags were often too expensive, with the goal of the ultra-low-cost tag still years from reality.[79] UHF RFID technology also did not always work as advertised. As I discuss in more detail in the chapter 3, tags often did not work well around water and metal, limiting the types of product pallets to which they could be attached. Consequently, Walmart's mandate was criticized for not recognizing technological limitations, for not acknowledging the significantly higher cost of RFID compared with barcodes, for moving too quickly, and for putting too many demands on manufacturers to maintain one RFID system for Walmart and one barcode system for everyone else.[80]

The Walmart mandate was, in the short term, a huge win for RFID tagging. Forecasting organizations called it RFID's tipping point, investors predicted RFID growth, and articles began predicting all kinds of future RFID uses, ranging from "smart" appliances to "active" prescription bottles to combat counterfeit drugs.[81] Walmart and Gillette also partnered to test "smart shelf" technology that would detect whenever someone moved a product from a shelf. For a technology used in the supply chain, RFID was receiving a lot of public attention.

In the longer term, however, the Walmart mandate was possibly one of the worst things that could have happened to RFID in the supply chain. Immediately following the 2003 mandate, a few markers looked good for RFID's growth, and sales of passive EPC tags increased from 2 million in 2003 to 120 million in 2005.[82] But growth stalled from there, and Walmart and other companies scaled back their mandates. By 2007, it was clear that RFID deployment in the supply chain would not meet the demands of the mandates. Walmart's mandate in particular was labeled as a failure.[83]

Then the situation got worse. RFID deployments in the supply chain did not achieve initial goals but were still increasing slowly until the financial crash of 2008. The 2008 crash led to a decrease in the amounts of money companies invested in new projects, and according to industry people I interviewed, many corporations saw RFID as an unnecessary expense because barcode systems were already in place. Multiple people I talked to described the period between 2008 and 2011 as the "dark days," when many RFID hardware and software companies failed. Some large companies did continue to experiment with RFID in the supply chain, but for the most part, from 2008 to 2012, RFID adoption stalled. The scope of the slowdown can be measured in a few ways, but two anecdotal measures are that readership of the flagship publication (*RFID Journal*) and attendance at the main industry conference (RFID Live) returned to pre-2008 levels only after 2012.[84]

The 2008 to 2012 period was likely the bottom of RFID's "trough of disillusionment."[85] Some companies halted pilot projects while others chose not to proceed with planned projects. Some RFID companies failed altogether. However, with the general uptick in the global economy, forecasts for RFID in the supply chain began to slowly improve after 2012, and the technology has now begun to meet some of the potential predicted in the early 2000s. To quote a 2018 study on adoption, "Despite a slower adoption than the hype suggested, the realisation that RFID can streamline apparel supply-chain management and bring transparency to the retail space, is generating a resurgence of interest in its potential."[86] Namely, some companies have moved past Walmart's initial goal of pallet-level tagging to return to Ashton's original vision of the Internet of Things: item-level tagging, which is basically the "holy grail" of deployment pointed to in industry publications.

After much disillusionment, item-level tagging has begun to pick up. Clothing companies such as Lululemon,[87] American Apparel,[88] and Marks & Spencer tag nearly every single item in their inventory. Marks & Spencer, for example, tagged over four hundred million items in 2014 alone.[89] Other companies such as Macy's, Bloomingdale's, Adidas, Tesco, Walmart, and many others now tag all items in parts of their supply chain. With just a fraction of the attention the technology received in the mid-2000s, RFID has now begun to reach some of the levels of ubiquity predicted at the peak of the technology's hype phase.[90] After the aborted launches of the early and mid-2000s, the last few years have seen increased optimism and a new round of articles with titles such as "Is the RFID Revolution Finally Here?"[91]

As a final point, even as RFID was being labeled a "failure" in various publications, it continued to cement its place as communicative infrastructure in areas outside the supply chain. Throughout the 2000s, the number of RFID toll roads expanded, RFID for animal identification increased and became mandated by some countries, more cities adopted RFID cards in public transportation, RFID school-systems matured, RFID's presence in access control systems increased, RFID chips began to be required in ePassports, and contactless RFID-enabled credit cards spread throughout Europe and Asia. These successes, however, were often overshadowed by the struggles of item-level EPC tagging, struggles that have recently begun to be overcome. And while the struggles are important to note, so are the successes. As I argue in chapter 4, much of the discussion of the coming Internet of Things focuses on a relatively small range of uses, whether for item-level tagging or networked domestic objects. The focus tends to ignore the many examples of communicative infrastructure, ranging from toll roads to payment systems, that already animate the physical environment.

Conclusion

A few strands unite the technologies discussed in this chapter. For one, except for maybe QR codes, the technologies I covered mostly remain outside the typical purview of social research. Barcodes are a taken-for-granted technology of retail and the supply chain, IP addresses are mainly a topic for engineers and remain all but unnoticed by most people using the internet, radar is a logistical and military technology, and RFID is an oft-ignored piece of various infrastructures of mobility. But more importantly, what these technologies share is their role as infrastructures of identification. Whether the identified object is a computer on a network or a plane flying through the air, the technologies covered above differentiate objects in the physical world.

Each of these technologies has a unique history that could easily take up a chapter on its own. Rather than go too deeply into each history, I have instead provided background useful for understanding how RFID differentiates objects within wide-ranging networks of communication. Each technology provides a different piece necessary for understanding the potential impacts of RFID: barcodes are RFID's most important technological antecedent and are still the most widely used AIDC technology; QR codes foreground the communicative potential of NFC as RFID becomes linked to smartphones; IPv6 shows the necessity of expanded identification capabilities of the coming Internet of Things; radar developments in backscatter communication were necessary for the growth of passive RFID. These related technologies of identification all play a role in the past, present, and future of RFID development.

Possibly most concretely, the growth of both RFID and IPv6 illustrates a larger trend traced throughout this book: the increasingly granular networks of identification that sort objects and bodies in the physical world. RFID and IPv6 are technically quite different and are not reliant on each other. But they each have been identified as a way to vastly expand identification capabilities. IPv4 just does not have the capacity to differentiate tens of billions of internet-connected objects. Barcodes do not have the capacity to assign unique codes to tens of billions of objects. By no means am I arguing that IPv6 or RFID have followed a straightforward, natural evolution from their antecedents. One could make that argument with IPv6, but it would be a difficult case to make with barcodes and RFID. For one thing, barcodes are still a dominant technology, and depending on when one labels the "birth" of RFID, barcodes are debatably the younger technology. But while few technological shifts follow a predetermined straight line, the increasing use of both RFID and IPv6 does represent a shift that underlies the core argument of this book.

Ultimately, this book is about identification, about how various infrastructural technologies can turn objects into data and differentiate things in the physical world. Identification in itself is often not at the forefront of discussions of technology, but much of our built environment relies on processes of differentiation.

Many grocery stores cannot function if the scanner can no longer tell a six-pack of Coke from a jar of peanut butter. My internet browsing does not work if my information request cannot be linked back to my computer. The practices of identification enabled through RFID contribute to everything from religious pilgrimages to pharmaceutical anti-counterfeiting, as detailed in chapters 4 and 5. But first, chapter 3 delves further into processes of identification by examining just how RFID works and the differences among types of RFID technologies.

3 Understanding RFID Technologies

One of the struggles I faced when I first started researching RFID was figuring out just what RFID is. Popular and academic sources sometimes confuse terms when discussing the technology, tending to define RFID in limited terms and ignoring that RFID includes a fairly broad suite of technologies that have different technical capabilities. Take the difference between an RFID-enabled credit card and an active RFID transponder as an example. The credit card has no internal power source and a read range of a few centimeters; the active transponder requires a battery and can be read from as far as one hundred meters away. Conflating the two is like confusing an intercom system with a mobile phone. Those conflations lead to confusion about what RFID is, what RFID can do, and what people should worry about when they worry about RFID.

Understanding the different types of RFID is an important step toward grasping their potential as communicative mobile infrastructure. One does not have to understand the physics behind backscatter modulation (I certainly do not) or the anti-collision protocols that enable readers to handle multiple tags at once (I, once again, do not). But knowing the difference between passive and active tags, high frequency and ultrahigh-frequency RFID, and tags that contain detailed data versus tags that contain numbers to grant access to that data, is crucial to understanding the varied ways RFID works as identification infrastructure.

Equally importantly, knowing the basics of how RFID works can help separate fact from fiction. In the mid-2000s when the backlash against privacy issues related to RFID was at its height, some commentators conflated types of RFID in misleading ways. In some particularly dire privacy warnings, RFID tags that could be read only from a few centimeters were lumped in with tags that could be read from many meters away. Tags that were encrypted to protect sensitive information were lumped in with supply-chain tags that were not encrypted to keep down costs. The differences matter, especially when discussing the affordances and constraints of RFID systems. Even now, the occasional conspiracy theory about governments injecting RFID and tracking citizens appears on a site and gets shared, ignoring that injectable RFID typically has a read range of less than an inch. The confusion even reaches a point that some trade organizations deny using RFID when they clearly do.

Ultimately, this book is not an engineering textbook. This chapter does not contain mathematical formulas or snippets of coding languages. I keep the technical discussions to a minimum, in part to make the chapter more readable and in part because I am not a physicist or an engineer. Consequently, the rest of this chapter goes into nontechnical detail on a few of the important aspects of how RFID works, the different types of RFID tags, and near-field communication (NFC). RFID is not just a short-range technology used in payment and transportation systems; RFID is not just a longer-range technology used in the supply chain. RFID is both of those things, and the capabilities of different systems shape the role the technology plays as an infrastructure of identification.

RFID Systems

When people talk about communication media, they tend to boil complex systems down to one device. As a mobile media researcher, I have written fairly extensively about smartphones. I have researched how people use smartphones to coordinate with others, look up information about their surroundings, and share information with their wider social networks.[1] In almost all my research, I talk about smartphones as if they are stand-alone technologies that operate on their own, a practical choice that lets the analysis focus on the level of the human-centric interface. In reality, however, when people use a smartphone to look up mapping information, they interact with much more than a mobile device. The process involves a smartphone—and even saying "smartphone" ignores different pieces of hardware—that accesses infrastructure to achieve a mobile data connection, beams a signal to the satellites of the Global Positioning System (GPS) orbiting the earth, likely supplements the GPS location with location data from Wi-Fi routers and cellular towers, and so on. Without the larger infrastructural network, the smartphone is just an expensive iPod Touch with the Wi-Fi turned off.

RFID also consists of multiple pieces necessary for a working system.[2] A basic RFID system needs three pieces to work: a transponder, a reader, and software that connects the reader to a larger system.[3] The transponder is the chip or the tag that houses the data. As discussed later in this chapter, transponders come in various shapes and sizes and work at different frequencies. The reader is what activates the transponder in a passive system or receives the transmitted data in an active system. Readers can be mobile and handheld or fixed as in subway terminals or tollbooths. Then software—also known as middleware—often has to transmit and filter the data collected by the reader to a larger system.[4] For example, when someone drives through a toll station or scans inventory on a store floor, the reader does not host the full system that logs the data. The data is instead transmitted to a larger database.

A relatively straightforward access control terminal can illustrate how one type of system works at a basic level. Someone trying to enter a secure building has an RFID-enabled badge (the transponder). The person waves the badge in front of the reader, which sends out radio waves to power the RFID tag and receives the data on the badge. The middleware authenticates the data with a database. If the data contained on the tag authenticates, the door to the room will open. That basic description also applies to similar systems like contactless payment terminals, but other RFID systems may not require any response. For example, an inventory check of RFID tags or a scan to collect sensor data might log the data without doing anything immediate. Those systems still work on the same basic principle, and a reader interacts with the transponder, collects the data, and shares that data with some kind of database.

The description above is basic and rudimentary, and RFID systems are complex and can be difficult to set up. In addition, the capabilities of the transponders and readers vary depending on a host of factors. The sections that follow go into more detail on types of RFID, particularly focusing on how tags work and how frequencies enable altered forms of object communication. As discussed earlier, knowing the difference between types of tags and RFID frequencies is an important step toward avoiding misunderstandings about just what RFID can do as mobile infrastructure.

Passive versus Active

The ubiquity of RFID tags is a major part of what makes the technology so interesting. Billions of tags are affixed to or embedded in objects each year, and that number continues to increase.[5] A major part of that ubiquity can be explained by the three ways RFID tags are powered: passively, semi-passively, and actively. Passive tags are by far the most common form of RFID, so it makes sense to begin, first, with passive tags.

Passive tags have no internal power source. They have no battery to power the antenna. They sit in an "off" state, waiting to be activated when they come in contact with an electromagnetic field emitted by an RFID reader: "The tag gathers energy from the reader antenna, and the microchip uses the energy to change the load on the antenna and reflect back an altered signal."[6] At that point, the passive tag is in an "on" state as long as it is within range of the reader. Once the tag leaves the reader's field, it returns to its passive state, waiting to be activated again by a reader.

The lack of internal power source is part of what makes RFID one of the key contemporary infrastructures of identification, for three primary reasons: (1) Passive tags can last a very long time. (2) Passive tags are cheap. And (3) Passive tags can be as small as a grain of rice.

Passive tags last a long time because they have no battery that needs to be replaced. Ostensibly, as long as the pieces of the tag remain in place, they can be powered by a reader. The lack of internal battery makes passive tags ideal for systems in which replacing batteries may be impossible or costly, and that design is why an RFID-enabled subway card can still work a decade later.

Another important point about passive tags is their cost. Batteries cost money, so tags that have no internal power source are cheaper. Some passive tags, when bought in bulk, now cost less than ten cents a tag, though price depends on various factors. The low cost has contributed to RFID's ubiquity. Billions of passive tags can be attached to low-cost objects and communicate data without the line-of-sight required by barcodes.

Relatedly, the lack of a battery means that passive tags can also be quite small. The image in chapter 1 (figure 1.1) of tags I collected shows how diverse these tags can be, ranging from larger tags with longer read ranges to tags that are only a few centimeters in size. Tags can also be nearly two-dimensional, with the antennae found inside some smart cards measuring less than a centimeter in depth. Size is what enables passive tags to remain hidden, such as instances where the tags are embedded in cardboard or sewn inside pieces of clothing. All passive tags share the lack of an internal power source. However, as explained later, passive tags also work at different frequencies, and the frequency, along with the antenna design and power of the RFID reader, determines the tag's read range.

In a second type of tag—battery-assisted passive (BAP) tags—read range can be boosted through the semi-passive middle ground between passive and active. BAP tags have batteries that power the chip's circuitry, but they communicate only when drawing power from an RFID reader. Often, the battery is used to either boost the tag's read range or power additional hardware such as a sensor attached to the tag. The disadvantage of BAP tags are that they have a battery that will eventually need to be replaced, and they cost significantly more than passive tags. Consequently, they are used primarily in certain types of sensor arrays and to track higher-cost items.

The third broad type of tag is active RFID, which has an internal power source. These tags have the power to broadcast their signal constantly without the help of a reader, and the tags also may contain more internal processing power than passive tags. Unlike BAP tags, active tags do not require the assistance of a reader to broadcast. Active tags are necessary in a few situations: for example, in complex sensor arrays in which sensors take constant readings and need to transmit that information consistently and log the data. Another common use of active RFID is to track high-cost items through the supply chain.

The advantages of active RFID are that the technology can contain more complex computing power, can be more accurate at capturing reads, can log data, and can be read at longer ranges. Some active tags can be read from as far as one hundred meters away, making them ideal for tracking items in large spaces. Whereas passive or semi-passive systems can be difficult to set up because readers have to be placed strategically to power tags, active systems require fewer readers and will broadcast even if a link is not made with the reader.

With those advantages come disadvantages. First, active systems require occasional maintenance because they have batteries that need to be replaced. Second, active tags tend to be much bulkier than most passive tags because they have internal power. Third, active tags tend to be more expensive than passive tags and can cost anywhere from $15 to $100 depending on the tag.[7] Because of their maintenance, size, and cost, active tags are not an option for tracking low-cost items or being used in most kinds of transportation infrastructure.

Most of the RFID examples I discuss in this book are of the passive variety. Passive RFID is far more ubiquitous, with billions of passive RFID tags attached to objects and embedded in everything from wristbands to passports. However, what I hope this discussion has made clear is that there is no "better" form of RFID. Passive is more appropriate for most applications, but BAP and active tags are superior for tracking high-cost objects in real time or powering some sensor arrays. The repeated refrain of "it depends" when comparing RFID technologies also extends to the comparison of the different frequencies at which RFID operates. As the next section explores, whether a tag has internal power or not is only one factor in determining measures such as read range and accuracy. Just as important is the frequency at which the system operates, a point that has led to some confusion in discussions about the social impacts of the technology.

The Agencies of Frequencies

The first category of frequency is low-frequency (LF) RFID, which refers to the International Telecommunication Union (ITU) designation for radio frequencies between 30 kHz and 300 kHz.[8] The advantage of lower frequencies is that the radio signal is not as sensitive to interference and can be read more easily around water and metal. The disadvantage tends to be that LF RFID transmits data more slowly than other frequencies and has a short read range of around ten centimeters, though, as with all RFID systems, the read range is also determined by the size of the antenna in the tag and the signal from the reader. While LF was one of the earliest widely adopted forms of RFID (mostly in animal tagging) and is still used for some microchips in pets, the technology fell out of favor, in many cases, as people turned to high-frequency (HF) RFID.[9]

The HF band of the radio spectrum falls between 3 MHz and 30 MHz. HF RFID—which includes near-field communication (NFC)—most often operates at 13.56 MHz. HF RFID tags have read ranges between ten centimeters and one meter (most often the shorter end), depending on the design of the system, and they typically transfer data faster than LF applications. They also are more sensitive to interference from metal and water, though HF does not have problems in most cases because the read range is not that long. HF RFID is used in a diverse array of applications, including contactless payment cards, passports, public transportation cards, some access control, the NFC found in many smartphones, most RFID human implants, and some supply-chain uses. An advantage of HF RFID is that the relatively limited read range is long enough for people to wave a card or tagged item in front of a reader, but not so long that the signal becomes disrupted or read by accident. HF is also less susceptible to environmental interference than ultrahigh frequency (UHF). In addition, HF RFID has strong standardization from the International Organization for Standardization (ISO), with standards covering everything from HF payment protocols to the use of EPC on HF tags.

The UHF band is designated for radio frequencies between 300 MHz and 3 GHz. UHF RFID most often operates between 860 MHz and 960 MHz. The UHF band for RFID is more complicated than LF or HF because the exact frequency band UHF RFID is allowed to operate at depends on the region. For example, the EU RFID allocation ranges from 865 MHz to 868 MHz, North American from 902 MHz to 928 MHz, and Chinese from 840.25 MHz to 844.75 MHz and 920.25 MHz to 924.75 MHz.[10] To quote the *RFID Journal,* "Many other devices use the UHF spectrum, so it will take years for all governments to agree on a single UHF band for RFID."[11]

UHF RFID was first patented in the 1990s by IBM and became a source of excitement in the RFID industry for a few major reasons. For one, UHF RFID has longer read ranges than HF RFID. Depending on the design of the tag and the power the reader, passive UHF RFID can be read from twelve meters or more, with a typical system having a read range of around three meters. UHF tags are also easier to produce and cheaper than HF tags. The low cost of UHF passive inlays is what led to the hype in the mid-2000s about low-cost tags that would make item-level tagging economically feasible. Finally, UHF tags can transfer data at faster speeds than HF tags. The high data transfer speeds make it possible to move a large number of items quickly past a reader and capture accurate reads, making UHF tags particularly useful for item- and pallet-level tagging.

To picture why UHF tags are used for item tagging in the supply chain, imagine a warehouse that logs inventory. The products are tagged individually and are read from a distance of a few meters when they pass by a reader on the warehouse floor. The increased read range makes it possible to read items without placing them too close to the reader, and the faster data transfer speeds make it possible for a reader to log a large amount of data in a small amount of time. However, for all the benefits of passive UHF RFID, the technology has some drawbacks. For one, the higher the frequency range, the more sensitive radio signals are to interference. Consequently, UHF RFID signals can bounce off metal and be absorbed by water, making it more complicated to tag items that feature metal or water. Advances in tag design have partially solved the problems with UHF interference, but UHF tags remain more sensitive than HF tags. In addition, UHF systems can be more complex. Because the read range is longer, the signal is more easily dispersed. Readers have to be set up strategically to ensure items will be read, and systems have to take into account environmental factors.

As a final note on frequencies, within the broader UHF band, some systems operate at 2.45 GHz, which is the frequency used for Wi-Fi, microwave ovens, and Bluetooth. A few types of specialized RFID tags, both passive and active though mostly active, operate at 2.45 GHz and enable even longer read ranges. While active RFID is not as commonly used, the higher-frequency ranges of many active systems enable certain uses, such as large real-time locating systems (RTLSs) and the tracking of expensive goods across long distances.

Frequency is one of the most influential factors that shapes RFID systems and the one that is most confused in some popular accounts of RFID. The confusion often appears in the differences between HF and UHF RFID. UHF is used mostly for item tagging and longer read uses like tolling. The long read ranges and environmental sensitivity make the technology less than ideal in subway infrastructure or payment cards. However, some of the writings about the downsides of RFID confuse the technical capabilities of the two technologies. Fears about RFID in passports, credit cards, and so on sometimes talk about those forms of RFID as if they could easily be read from multiple meters away. Warnings about RFID implants envision *Minority Report*-style futures in which people would be identified at a distance through implants.[12] Although it may be technically possible to read HF RFID from multiple meters with super-strong readers, that scenario remains unlikely. RFID implants are an example why: People could implant UHF tags, but UHF does not work well around water. Human bodies are mostly water.

The distance from which a tag can be read affects how people talk about RFID as a technology, so it is important to know the basics of how RFID frequencies work. The different affordances of radio frequencies also illustrate the necessity for some precision when discussing "thingly" agency.[13] Radio waves are generated all around us. They power Wi-Fi connections, radio communication, various forms of RFID, and so on. But the human behaviors enabled or restricted through technologies that use these radio waves are shaped by the capabilities of different frequency levels. Consequently, an HF form of RFID exerts a different type of agency in the physical environment than a UHF form of RFID.

The HF form requires more intentionality and precision from the person using the technology. Humans have to physically place a tagged object within a few inches of a reader, or to put it differently, a human will likely play a more prominent role in a network of HF RFID. UHF RFID, on the other hand, can work more as an autonomous actor in a designed system. The tags can be read from a long distance and might require no intentionality from any human being manipulating the tagged object. For example, a student with a UHF-tagged bracelet may have their identity read without them even knowing. The difference in the agential abilities of different frequencies is a topic I return to in chapter 4. Ultimately, knowing the difference between frequencies provides a baseline of knowledge to explain why a car with a UHF toll tag can speed down a highway at seventy miles an hour while a contactless credit card will not work when more than a few centimeters from the terminal.

The Tags behind the Data

The final difference among RFID systems covered here involves types of tags. As already discussed, tags can be passive or active and communicate on different frequencies. But tags can also vary in complexity regarding memory and read-write capabilities. Passive tags, for example, can be read-only, read-write, or write once, read many (WORM). Read-only tags are generally programmed with a serial number when they are produced, and that information cannot be changed. Read-write tags can be reprogrammed when they come into contact with a reader, meaning serial numbers or identifying information can be erased and replaced with new information. WORM tags can be written to once and then become read-only. Of the three, the read-only tags are the cheapest, but they lack the flexibility of programmable read-write tags.

Beyond those three types, RFID tags can also vary in complexity. To some sources, RFID exclusively refers to the simplest types of tags that carry unencrypted basic information and are essentially just improved barcodes. Consequently, some people incorrectly restrict the RFID label to less complex tags that have no internal memory and cannot read or write data. For example, some sources differentiate between RFID and contactless payment technologies because those technologies have advanced security protocols, an integrated chip, and a larger memory capacity.[14] In the FAQ on their official site, the Secure Technology Alliance (previously the Smart Card Alliance) writes that smart cards are not forms of RFID. They then include the following explanation:

> The RFID technologies that are used to add value in manufacturing, shipping and object-related tracking operate over long ranges (e.g., 25 feet), were designed for that purpose alone and have minimal built-in support for security and privacy. Contactless smart cards, on the other hand, use RF technology, but, by design, operate at a short range (less than 4 inches) and can support the equivalent security capabilities of a contact smart card chip.[15]

What that paragraph actually says is that smart cards are not the same as passive UHF tags, not that smart cards are not RFID. RFID can have read-write capabilities and expanded memory, short read ranges, and enhanced security capabilities. A smart card can have a chip that stores data, but it still communicates through RFID and is powered by an RFID reader. Obviously, the RFID technology in a smart card and the increased power of the microchip will cost more than a read-only passive UHF tag, but that does not mean the technology is not RFID.[16]

Additionally, the data on an RFID tag can be varied and present different types of risk depending on the design. For example, studies have shown that some RFID ID cards (e.g., the national US Passport Card) can be skimmed by hackers who access the information.[17] That is obviously not ideal, but depending on the system design, the data being skimmed from the card is not personally identifiable information.[18] Rather, the data is a serial number used to access database information, which is different from having the social security number and biometric information coded into the chip. Another example is the data on RFID-enabled credit cards. Various sources have published videos of people "skimming" RFID cards to get the data,[19] and a large market in "RFID blocking wallets" has arisen to address those concerns. But despite the fears and alarming news stories, it is debatable how much of a concern RFID card skimming ever should have been in the first place.[20] While early RFID cards might have included credit card numbers, cards now host data that sends a one-time code for a transaction, meaning the data is not worth much to scammers. In addition, for all the hype around RFID card security, there have been few reported cases of people having their card numbers stolen through RFID skimming. A Snopes fact-check article pointed out that the Identity Theft Resource Center had "never seen a case of RFID skimming used to steal information."[21] The data on the card is most often encrypted, limited to purchases of less than $30 in many places, and has to be physically accessed by someone nearby. Not to be glib, but there are far easier ways to get stolen credit card information. RFID security matters, but when discussing the security of different types of tags, it is important to also reflect on the type of data on the tag.

To add an extra layer of complexity, RFID can also be paired with sensor technology that enables the communication of different types of data. In fact, the combination of RFID and sensors represents one of the more exciting areas of growth for RFID's role in the Internet of Things.[22] Passive, BAP, and active RFID can all be paired with sensors and have been used to monitor environmental factors such as temperature, humidity, light, and strain on pieces of infrastructure.[23] Active tags allow for much longer read ranges, and because they have an internal power source and more memory capability, they can work as data loggers for sensor data. Active RFID is particularly useful when comprehensive records of environmental conditions are necessary because the systems can transmit data in real time and log large amounts of data that can be retrieved later. BAP RFID can also log data, and the battery lasts longer because it is powered only when in contact with the reader. However, BAP systems typically do not produce sensing data that is as comprehensive because the sensors are powered by the battery only when in contact with the reader.

Battery-enabled RFID systems that power sensors are an established technology used in food and pharmaceutical monitoring, and they are also used to evaluate strain on infrastructure such as bridges and the humidity level of objects. However, possibly the most significant future use of sensors paired with RFID may come in the form of passively powered sensor technology. Passive systems can use the field emitted by the reader to provide power to sensors and then transmit environmental information. These systems generally have shorter read ranges, though UHF systems still have read ranges up to around five meters. They also cannot log data because they do not have the internal memory or power to record the data on the chip. But they cost less and do not require batteries that need to be replaced.

One major advantage of passive UHF sensor tags is that they are cheap. According to a 2017 study of RFID and sensors, the average BAP sensor tag costs $16, which is a limiting factor.[24] Passive UHF tags cost significantly less and have now passed BAP and active sensors in the RFID market. In 2017, around 330,000 BAP sensor tags were sold, compared with 5.2 million UHF sensor tags. The low cost and smaller size of passive sensor tags have opened up new opportunities for RFID sensing. For one, some manufacturers include passive sensing tags in pieces of construction to check for leaks and no longer have to remove them once a product is built. Another example involves UHF sensors that have been used inside the adult diapers in health-care facilities so that handheld readers can detect moisture rather than health-care staff having to manually check the diaper. Passive NFC (HF) sensor tags are also a growing market and have begun to outpace active and BAP RFID sensors. The forecasting numbers suggest the future of RFID's pairing with sensors will likely focus on the low-cost passive variety.

Passive UHF and NFC tags are quickly becoming an established part of the Internet of Things. However, possibly the most significant pairing of RFID and sensors is still a few years from realizing widespread commercial realization: RFID and wireless sensor networks (WSNs). As I discuss in chapter 4, the ability to sense environmental information on large scales will be key to the growth of the Internet of Things. Much of that sensing will be transmitted through WSNs, which

> can be defined as a network of tiny devices, called sensor nodes, which are spatially distributed and work cooperatively to communicate information gathered from the monitored field through wireless links. The data gathered by the different nodes is sent to a sink which either uses the data locally or is connected to other networks, for example, the Internet (through a gateway).[25]

WSNs consist of a large number of cheap sensors that monitor the environment. Once again, the sensing information can pertain to everything from temperature to physical strain on infrastructure, and the sensors produce data that can be used to elicit changes through software-controlled systems. As computer scientists Priyanka Riwal and colleagues argue, "It will not be an exaggeration to consider WSNs as one of the most researched areas in the last decade."[26]

Many sensor networks are powered by either deploying cable or using batteries to power sensors. Both designs face problems in terms of cost, maintenance, and requirements for extensive infrastructural development. Consequently, a major area of research in sensor networks has explored using passive RFID to power WSNs. In a comprehensive literature review focusing on passive RFID and sensors, Jun Zhang and colleagues found that between 2005 and 2015, 442 articles were published on the topic, though their review did include sensor networks passively powered in other ways.[27]

Passive RFID antennas combined with sensors have the potential to make wide-scale WSNs cheaper, easier to maintain, and simpler to deploy because the sensors can be used in already existing RFID systems. In these networks, RFID would wirelessly communicate the sensor data and make each deployed sensor uniquely identifiable. At this point, research has explored using passive RFID to power WSNs to monitor everything from agriculture to the temperature of concrete.[28] While the pairing of the two technologies is still in its relative infancy, the ability to passively power WSNs may end up being one of RFID's most substantial contributions to the future of the Internet of Things.

Near-Field Communication

RFID can work with or without internal batteries, operate at different frequencies, and include a variety of tags with different capabilities. In sum, RFID is varied and complex and is more than a technology used solely for logistical purposes. To complicate it all a bit more, I now want to move on to near-field communication (NFC). Some sources talk about "RFID *and* NFC,"[29] while other call the two "related technologies."[30] But the reality is simultaneously more straightforward and more complicated: NFC is a type of RFID that falls under the larger umbrella covered in this chapter. NFC technology was developed out of the ISO standards governing HF RFID, and NFC devices can read HF RFID tags that use the ISO 15693 standard.[31] So NFC is not different from RFID; it instead is a type of RFID, but one that, in a sense, "democratizes" RFID—for better or worse—by turning smartphones into both RFID tags and readers.

Because NFC is an HF technology and is based on ISO standards governing proximity cards, its read range is only a few centimeters, an intentional choice that enables increased security and fewer accidental reads. But two differences between NFC and other forms of RFID make the technology potentially so significant: (1) most smartphones have NFC capabilities (see figure 3.1), and (2) NFC-enabled devices are bidirectional and can operate as both tag and reader. As an example, a smartphone works like an RFID tag in contactless payments systems like Google Pay or Apple Pay. People can wave their smartphones in front of a payment terminal just as they could do with a contactless RFID payment card. But in other examples, such as using a smartphone to access an NFC tag in a poster, a smartphone acts as a reader. For a hands-on example, people can try to read a subway card or other

Figure 3.1
A disassembled Nexus 5 Android phone.
The white arrow points to the phone's NFC controller. Photo courtesy of the author.

HF RFID tags. If the tags are NFC-compliant, the smartphone can pull data off the card. Because of the dual capabilities, NFC can also be used to pair devices without requiring the Bluetooth pairing process.

NFC has been heralded as a transformative mobile technology for more than a decade. It has the potential to put RFID capabilities into the hands of smartphone users and enable new ways to link smartphones with infrastructure. However, for all of its potential, at points NFC has seemed like it was going to be a failure, so it helps to spend some time on its history.

NFC first gained attention in 2003 when it was approved as an ISO standard. The original press release described NFC as "a very short-range protocol, for distances measured in centimetres, and is optimised for intuitive easy and secure communications between various devices without user configuration."[32] Upon its release, NFC was met with immediate hype. For example, in that initial press release, Philips Semiconductors was quoted as saying that "Philips envisions a world where everyone can always connect to information, entertainment and services with NFC playing a vital role in making this happen."[33]

Philips was not alone in its optimism about NFC. In 2004, Nokia, Sony, and Philips partnered to create the NFC Forum.[34] Despite the backing of an influential trade group, however, NFC was not widely included in devices throughout the 2000s. As described in a 2008 article in the *RFID Journal*, "Since the technology was first announced in 2003, it has failed to win over any handset manufacturers to start mass production of NFC-enabled phones, or any mobile operators to buy such phones and promote NFC services to customers."[35] The GSM Association, one of the main standards organizations for mobile networks, made a public call in 2008 for NFC functionality to be built into mobile phones by 2009,[36] but that soft milestone was also missed. Throughout the latter half of the 2000s, it seemed as if NFC adoption might never happen.

NFC's fortunes improved in 2010 when Samsung released the Nexus S, which was the first Android phone to include NFC. In 2011, NFC support was included as part of the Symbian operating system, and more and more Android phones and BlackBerry phones were shipped with NFC. Mobile payment applications that relied on NFC were also released around this time. For example, Google Wallet (now Google Pay) was released on NFC-enabled Android phones in 2011.[37] Mastercard, Visa, and Square released NFC payment applications not long after, and Google also released Android Beam, which uses NFC to share data between two Android devices. NFC then gained more traction in mobile payment systems, with Apple announcing Apple Pay in 2014 and Samsung introducing Samsung Pay in 2015. The firm Strategy Analytics predicts that "the value of transactions conducted via NFC handsets will grow from US$30bn in 2016 to US$45bn in 2017, US$70bn in 2018, US$110bn in 2019, US$160bn in 2020 and US$240bn in 2021."[38]

All these mobile payment applications use NFC,[39] but that does not mean they always used NFC to communicate the same type of data. Google and Apple, for example, initially designed divergent approaches to NFC mobile payments. With Google Wallet, Google served as an intermediary that stored a person's credit card information on their servers. When someone used Google Wallet, their phone used NFC to transmit a Google virtual card to the merchant, but the merchant did not see the person's actual credit card information. Google then used the stored card information to charge the bank that issued the credit card. Apple took a different approach that used a process called tokenization. When someone uploads credit card information to Apple Pay, the system contacts the issuing bank and "upon confirmation receives a device and card specific token called the Device Account Number (DAN) that is stored on a secure chip on the device."[40] Apple Pay then transmits the token to the merchant and the payment is authorized with the bank, just like with other types of credit cards. Apple, unlike Google, did not host the credit card information and instead relied on the token rather than the credit card number.

The original Google Wallet system supported "all credit and debit cards from Visa, Mastercard, American Express, and Discover"[41] because Google did not need approval to store that information. However, with the switch to Android Pay and now Google Pay, Google turned away from the virtual wallet system to embrace the tokenization built into the original design of Apple Pay.[42] Tokenization offers increased security, but it also means that Google and Apple need to form partnerships with banks to receive the token because Google itself no longer serves as the mediator like it did with Google Wallet. Consequently, the business model for NFC payments has shifted. With Google Wallet, Google could keep some data hidden from banks and use the data themselves. With the token systems of Google Pay and Apple Pay, the bank collects the data, but they pay an unspecified transaction fee. Consequently, with a system like Apple Pay, the value to Apple is likely not through data collection, but rather through transaction fees from banks Apple made deals with.

Google, Apple, and a variety of other companies have been competing in the mobile payment market for years now. For most of the 2010s, however, NFC has been less successfully adopted in mobile applications that focus on areas other than payments. The lack of NFC adoption in other types of applications occurred in no small part because of decisions made by Apple. Apple is a major player in the mobile phone market and one of the world's largest companies by valuation; consequently, Apple makes decisions that affect far more than iPhones. For example, in 2010 Steve Jobs published a letter explaining why Apple would no longer allow Adobe's Flash platform on the iPhone. Developers who used Flash then did not have access to the iPhone's lucrative demographics, so it made little sense to use Flash in Android applications unless a developer wanted to design two significantly different applications. By 2012, Android also stopped supporting Flash.[43]

Apple's outsize role in the mobile phone market has had a similar, though less drastic, effect on NFC. Apple did not include NFC in iPhones until the iPhone 6 (released in September 2014), a few years after NFC became a regular feature in Android phones. Even after the release of the iPhone 6, Apple restricted developers' access to NFC, enabling only Apple Pay to access iPhones' NFC capabilities.[44] Apple claimed it restricted NFC because of security reasons,[45] though critics argued the explanations were misleading and the real reason was to protect Apple Pay from competitors.[46] Regardless of the reason, Apple's restrictions on NFC slowed the technology's development, particularly in mobile applications. Developers could include NFC capabilities only in Android applications, so they would have to eliminate that feature from an iPhone version. Equally importantly, it was often not financially viable for companies to use NFC tags in advertisements or product packaging when only a segment of smartphone users could access the tags. One research site I visited had stopped asking for participants in their NFC demonstrations because they had too many groups in which everyone owned an iPhone.

Apple's reluctance to enable NFC did not completely stop the spread of the technology. Companies have developed interesting uses of NFC tags to link smartphones to objects and bridge gaps between physical objects and digital information. For example, as mentioned in chapter 2, some marketers have used NFC tags to replace QR codes in print materials, enabling people to tap a phone against a poster to be linked to a website or video.[47] NFC has advantages because it does not require specialized applications like QR codes or the pairing process of Bluetooth, and anyone who has NFC activated in an Android phone can access a tag to go to the digital endpoint. In addition, NFC does not require an optical line of sight like QR codes do, so tags can be embedded inside materials. One example is the Dyne Menswear clothing line that includes NFC tags inside their clothing. The tags link people to everything from information about where the clothes were produced to song playlists associated with the piece of clothing.[48]

Another growing use of NFC tags is to combat counterfeiting. Millions of products are counterfeited each year, and some companies have used NFC tags for authentication. In these cases, the NFC tags work as less-robust versions of Sterling's spimes, discussed in chapter 1. They communicate part of their history with end users. Here are a few examples:

- The European fashion company Moncler provides buyers with the option of downloading an app, reading the tag on a piece of clothing, and receiving product authentication from the company's website.[49]
- Winemakers have attached NFC tags to bottles people can scan to ensure the wine is not counterfeit and is the properly labeled vintage.[50]
- High-end products, such as Johnnie Walker Blue Label, have used NFC Thinfilm to let users check authenticity and determine whether the seal on a bottle has been broken.[51]

Those are only a few examples of companies that have used NFC as a bridge between physical objects and digital information. Of course, the companies also potentially benefit from customer interaction with NFC tags. Beyond the obvious goals of driving sales with product information, companies that have people use a specific app to access the tags can potentially track post-purchase behaviors in new ways. If a company were to design an app and have consumers scan tags before breaking the seal on the product, the company could be able to track when products are opened and potentially make links between when individual products are purchased and when they are consumed. NFC tags that operate as pseudo-spimes leave a digital trace that could increase companies' ability to track behaviors outside store settings.

Until recently, companies that use these tags could interact only with non-iPhone users who could access the NFC capabilities in their phones. That limitation, however, was partially lifted in June 2017. Apple announced it would remove some restrictions on NFC in iOS 11, enabling the iPhone to read non-Apple NFC tags.[52] Even with the lifting of restrictions, iPhones still do not have all the NFC functionality of Android phones. For one, iPhones cannot write and program NFC tags, unlike Android phones that have "write" functionality if someone downloads the correct app (e.g., NFC TagWriter by NXP). In addition, iOS11 does not offer native support for NFC tags. Android phones will automatically read an NFC tag if NFC is turned on. With iOS11, the phone needs to be running an app that uses NFC features; tags cannot be read from the home screen.

Even with those limitations, Apple's decision led to excitement in the NFC industry and will possibly contribute to a fairly rapid expansion of applications that use NFC to connect phones with the physical environment.[53] Companies will also likely increasingly tag high-end products with NFC to provide increased interactivity after people purchase a product, though they will have to run certain apps to access the tags with an iPhone. As of this writing—less than a year after Apple's announcement—it is still too early to tell how far NFC will spread, but the future for the technology looks brighter than ever.

Conclusion

When technologies work like they are supposed to, people rarely think about how they make things happen. I rarely reflect on cell towers unless my phone cannot get a signal. I do not think about GPS satellites unless I cannot get my location. Technologies typically go through a process of "black-boxing," which philosopher Bruno Latour defines as "the way scientific and technical work is made invisible by its own success."[54] RFID is an example of that type of black-boxing, and when RFID works as it is supposed to, many people likely do not even know the technology is there. Until I began my research for this book and started interviewing people and reading technical manuals, I had no idea the same suite of technologies got me through a subway turnstile, unlocked the office building I used at a previous job, linked my car to a toll system, and let me make contactless payments. When I did unknowingly use RFID, things happened as if by magic. I never stopped to ask why.

Part of the infrastructural imagination discussed in chapter 1 involves learning about RFID and then researching whether a system used RFID or not. In many cases, such as researching some sensor arrays, I was never able to figure out whether RFID was involved. The technical specifications were not available or were so murky that the data transmission and identification could have used some other technology. But in many cases, I began spotting what to look for when I searched out RFID. I looked for contactless systems that involved identification, most often serial numbers, as in the case of toll system and transportation systems. I adopted Ian Bogost's call to "make things" and bought an RFID reader and some tags to see how they worked.[55] I broke a subway card, a toll tag, an old smartphone, and an access badge to find the RFID antennae. I had students in one of my classes program NFC tags for a small assignment. But I also began searching out RFID in diverse locations. I went to stores I knew used RFID to find the UHF tags I read about. I visited a distribution center and got a tour to see how the site used UHF tags for inventory. I attended a tradeshow to see what kinds of tags salespeople would try to sell me. I dealt with their sad faces when I told them I was a professor who did not have any money.

Learning more about different types of RFID and the many places the technology can be found was not just an academic exercise. As mentioned in the beginning of this chapter, one of the major struggles of understanding RFID from a social perspective is first understanding what the technology can do from a technical perspective. Some articles confuse types of RFID to hype or downplay fears about privacy, limit the RFID label to only one type of the technology (most often UHF tags used in the supply chain), or outright label RFID technologies as something else. But as this chapter suggests, simply saying something is "RFID" often says little about the agential effects of the technology. Take the difference between passive and active tags. Passive tags can cost as little as ten cents and exist in near-perpetuity because they have no battery that needs to be replaced. They can be attached to or embedded in billions of items and transmit identifying information whenever they come in contact with an appropriate reader. Active tags, on the other hand, can contain more complex internal memory, transmit from as far as a hundred meters, and accurately track high-value items from a long distance. But they cost far more and require regular maintenance. The difference there is not minor.

By no means am I arguing that one has to become an expert on RFID to understand the role the technology plays as an infrastructure of identification. The main argument of this chapter has been that a baseline level of knowledge about the different types of RFID is fairly necessary for discussing the technology in any depth. An example I cover in more detail in chapter 4 that gained widespread attention shows why: In the summer of 2017, a Wisconsin tech start-up called Three Square Market began an optional program to inject employees with an RFID chip. The employees could then use the chip for access control and payment in the office cafeteria. Microchips are highly controversial, and they should be. However, when the story got national attention, the controversy was often discussed in inaccurate terms. Most RFID implants are HF RFID that can be read from a few centimeters, but some sources talked about the implants as if they could be read from meters away, confusing HF implants with the UHF tags typically used in the supply chain. Those differences matter.

Chapter 4 and chapter 5 go into more detail on how RFID has been used to connect and produce identification data about objects. When relevant, I point out the type of RFID used in various systems to give the reader a better idea of just how the system works and what its limits are. After all, as this chapter has shown, just saying something includes RFID does not mean much. However, despite the differences, RFID technologies share the affordances that shape the key argument of this book: they are examples of mobile infrastructures of identification that have the ability to turn almost any object into identifiable data.

4 RFID and the Internet of Things

Few movies now seem more dated than 1992's *Lawnmower Man.* The movie was never very good, but its focus on digital bodies wandering through a virtual reality world captured a popular view of how the internet would evolve, a view the last two decades proved to be rather misguided. With visions of the future shaped in no small part by films such as *Lawnmower Man* and far more acclaimed science fiction novels such as William Gibson's *Neuromancer* and Neal Stephenson's *Snow Crash,* people in the 1990s often talked about the internet as a space separate from the physical world. The online world was a "cyberspace," an "information super highway" that would negate spatial difference and make it so that, to quote a famous MCI commercial, "there would be no more there. There would only be here."[1] Much early academic thought about the internet took a similar perspective. Technology researcher Nicholas Negroponte wrote about how people would no longer need to socialize in physical spaces. Philosopher Paul Virilio talked about how individuals might soon stop having sex in the physical world.[2] Geographers predicted the "death of cities."[3] Online environments were often viewed as a way to live life through an avatar and ignore the corporeal constraints of the physical world. To summarize Negroponte, life was breaking down into a battle between a "world of bits" and a "world of atoms,"[4] and the bits were winning.

More than two decades later, the idea of a separate, dichotomized world of bits and a world of atoms seems like a relic from an earlier time. The internet did not replace people's need or desire to live a physical life.[5] The digital world did not supersede the physical world. In fact, there is no digital world and physical world; there is only one world. The digital relies on massive amounts of physical infrastructure to exist.[6] Physical bodies feel anguish when someone bullies them online. The internet is real, it is physical, and it materially affects the built environment. Go look for a CD store or read about what happens to frantic human bodies when the internet goes down at a major stock exchange. However, despite a partial move away from what sociologist Nancy Baym calls the "myth of cyberspace,"[7] people have still not fully abandoned the rhetoric of separation and division between the physical and digital worlds. Every time someone uses a phrase like, "In real life," they imagine a dichotomy between the "real" (the physical) and the "virtual/digital" (anything that happens online, I guess), as if a relationship that ends through a digital message is somehow less "real."

So why start a chapter about RFID and the Internet of Things by going all the way back to the virtual reality dreams of the 1980s and 1990s? The main reason is to reflect that, despite a lack of attention to materiality that still survives in some forms today, the internet was always made of "things," whether those things were computers, routers, cables, servers, or even the people on the other ends of screens. Discussions of people leaving their bodies behind to live their lives online often ignored that materiality, but the internet "as things" has always existed.[8] However, while the internet always relied on materiality, the newer conceptualizations of the Internet of Things represents a significant shift in how networked communication has become embedded in physical spaces. As shown throughout this chapter, the Internet of Things promises to extend computing to new spheres of life and connect things on an unprecedented scale.

The Internet of Things is a messy term that refuses simple definition, in part because it has almost become more of a marketing buzzword than an actual descriptor. But through all the hype, the Internet of Things describes a straightforward idea when boiled down to its basics: "the Internet of Things quite literally means "things" or "objects" that connect to the internet—and each other."[9] And importantly, one of the main arguments of this chapter is that the Internet of Things is already here and has been for quite some time. While many discussions of the Internet of Things focus on still-emerging technologies, the discussions in this chapter show that physical environments are already filled with diverse examples of object communication enabled through RFID.

The Internet of Things includes billions of objects that rely on various hardware and software. The strictest definitions require an object to have its own unique IP address to count as part of the Internet of Things,[10] but those definitions are too limiting and ignore the history of the concept. As discussed earlier, the term "Internet of Things" traces back to Kevin Ashton's 1999 presentation to Proctor & Gamble executives about RFID. Ashton used the phrase to describe using item-level RFID tags to link billions of objects. Notably, these objects would not have their own unique IP addresses; that cost too much then and still costs too much now in many cases. Instead, RFID tags could make each object uniquely identifiable and communicate with readers that would connect the data to the larger internet or a more contained intranet. More than fifteen years later, RFID is still one of the major enabling technologies of the Internet of Things, providing communicative agency to objects that often do not have their own computing power.[11]

As mentioned above, the original conceptualization of the Internet of Things focused on RFID tags and traditional readers, but the concept has evolved. Increasingly, near-field communication (NFC) is a major animating technology of the Internet of Things because NFC can connect objects to humans through smartphone interfaces. And many examples of Internet of Things technologies do not feature RFID at all, instead connecting objects directly to the internet. For example, one of the most popular Internet of Things technologies is the "smart" thermostat people can control from their smartphones. These devices have their own wireless internet connection. However, I am using the term more broadly to expand past more obvious examples of Wi-Fi-enabled "smart" objects to include the many things that are too small, too cheap, or too plentiful to have their own internet connection. That move follows Steven Weber and Richmond Y. Wong's urging that "the 'things' that Internet of Things will connect subsume and go beyond devices with computational capabilities, to include any and potentially all devices that have some ability to sense their environment or generate data about their interactions with other devices and/or people."[12] In many cases, these objects communicate through RFID, and in some cases they may not even communicate with the internet at all, at least not initially. Instead, they are able to communicate with the built environment in examples as mundane as a subway card talking to a terminal or as experimental as sensor technology combined with RFID that communicates information about ethanol levels in food.

The communicative function of RFID is the prime focus of this chapter. After all, the technology's role in the Internet of Things enrolls objects that may not have any computing power in networks of wireless communication. More subtly, however, the different cases covered below implicate key issues of identification. For the Internet of Things to function, whether through RFID, IP addresses, or however else, each connected object will most likely be uniquely identifiable. The conception of objects as "mass-produced" changes as objects become connected to infrastructure and to one another because each object holds unique identifying data in the attached or embedded RFID tag. For as much as discussion of the Internet of Things focuses on communicative objects, the processes of identification are equally significant. When objects become communicative through infrastructures of identification such as RFID, no two plastic cards, no two wristbands, no two sweaters with RFID tags sewn into the wool are the same. In the mass of nearly identical objects moving through the world, RFID plays a role in sorting similarity, classifying bodies and things, and tracking their mobility through various processes of differentiation. The entire conceptualization of "mass" in late capitalism may need to be rethought as uniqueness becomes a key part of how objects function and connect, even if that uniqueness is carried solely in a string of data on a tiny RFID tag.

The rest of this chapter examines the role RFID plays in different types of object communication. I split the sections among well-established examples—such as toll systems, subway terminals, and item-level tagging—and more futuristic discussions of RFID and the Internet of Things that include smart homes and biohacking. Ultimately, the examples are linked by the importance of three parallel developments: (1) increasingly complex practices of mobility; (2) progressively more intricate intertwining of software, hardware, and physical space; and (3) significantly more granular practices of identification. Consequently, before moving on to examples of RFID and the Internet of Things, I first discuss the role hardware and software play in the functioning of our physical spaces. I then examine different areas of everyday life in which one can already find objects communicating with one another. The Internet of Things is often positioned in contemporary discourses as something just off in the near future and about to happen, but if people look closely they can already find a built environment ready to listen to and sense billions of little things.

Technology and the Production of Space

Physical spaces are made up of technologies: roads, buildings, signage, and so forth. Many famous examples of how technologies shape place can be found in the literature, including how Georges-Eugène Haussmann rebuilt parts of Paris to create wide boulevards that made protest more difficult or how the extension plans of nineteenth-century Barcelona codified power in the gridded street plans.[13] Many other dramatic examples of how technologies shape places exist today, ranging from spikes set up to stop the homeless from sleeping to road systems that bulldoze through neighborhoods to emphasize cars over pedestrians. But many other ways that technology plays a role in what French philosopher Henri Lefebvre calls the "social production of space" are much more subtle.[14] After all, a physical place without technology is a natural landscape. To be lived in, a place requires technologies ranging from walls for shelter to water systems for sustenance. In other words, to return to Lefebvre's terms, how places are structured and how places then structure social practices and symbolic meanings cannot be separate from the various agential effects of technologies. Geographer Yi Fu Tuan wrote that "space is transformed into place as it acquires definition and meaning,"[15] but he could have just as easily said that space is transformed into place as it acquires technologies that shape meaning and make meaning durable.

Consequently, the idea that technologies exert agency over how spaces function is nothing new. The more recent significant shift, however, comes in how various technologies are able to communicate with the built environment through mediators such as RFID. The agency technology plays in the social construction of space then becomes more networked and distributed. These types of communication often go unnoticed, or at least unremarked-upon, but they are a crucial piece of how contemporary spaces operate. In no small part, the object communication that occurs in the background of interactions with the physical world shapes what bodies can and cannot do. It does so through what geographer Nigel Thrift calls the technological unconscious, "whose content is the bending of bodies-with-environments to a specific set of addresses without the benefit of any cognitive inputs."[16] In other words, decisions made about what humans are allowed to do are increasingly not made fully by humans themselves, at least in that moment; they are made by computing systems that partially govern how physical spaces function.[17] As a key identification and communication technology of the Internet of Things, RFID provides a significant amount of the data that informs the calculations that shape everyday encounters, which is why Thrift identified RFID as one of potentially the most important pieces of the technological unconscious.

When people interact with the physical environment, many of their actions depend on calculations of which they are not aware.[18] If someone uses a credit card—whether an RFID-enabled contactless card or not—the payment system authenticates the card. When a person passes through a toll, the toll assemblage authenticates a serial number through a database. When an employee at a store scans an incoming shipment to log the inventory, if the system does not recognize the tags, the inventory does not get logged. When these calculations do not work properly, people are not allowed to act in the same way. If the systems break down, the space often will not function as intended.

In all these examples, some kind of software and hardware exerts agency over how a physical space functions. If the technology does not work, the space does not work, or at least works in less than ideal form. The spaces become what researchers Rob Kitchin and Martin Dodge call "code/spaces" and "coded spaces":

> Coded space is a space that uses software in its production, but where code is not essential to its production (code simply makes the production more efficient or productive). Code/space, in contrast, is a space *dependent* on software for its production—without code that space will not function as intended, with processes failing as there are no manual alternatives (or the legacy "fall-back" procedures are unable to handle material flows, which means the process then fails owing to congestion).[19]

Kitchin and Dodge do more than show how the digital and physical have merged; instead, they show how the digital—through software—is often essential to how physical spaces operate.[20] This book looks at RFID mostly as hardware, as a material instantiation of communicative, mobile infrastructure, but discussions coming out of software studies are just as relevant. After all, RFID as hardware does not do much without software as a mediating layer that makes decisions about RFID inputs.

To see how, one can look at a relatively simple example of how RFID can be used to create code/spaces: RFID access control systems. Many buildings or rooms inside buildings require people to wave an access badge in front of a reader to gain entry (many contactless hotel cards work the same way). For example, in a former job, my building required an RFID-enabled key fob to enter on the weekends. If the software that read and checked the ID of my key fob did not work, the building essentially stopped being an office building on the weekend. No one gave us the option of a physical key. The building became an impenetrable set of walls, empty and nonfunctioning until the software or hardware was fixed. And the same is true of many other code/spaces involving RFID. A subway terminal that relies on RFID readers stops being an access point if the reader stops working. A distribution center cannot log inventory if the readers stop working.

Ultimately, theories like code/space examine the links between spatial experience and technology. Some of those links can seem rather obvious. Of course technologies, in the forms of roads, buildings, signage, public transportation, and so on, compose key pieces of the built environment. But the significant shift captured through concepts such as code/space is how spaces are increasingly mediated by computing, often networked computing that feeds back into the internet (or an intranet). Many actions bodies take every day are allowed or disallowed by calculations that remain out of sight and mostly out of mind.

The discussion in this chapter builds on the concept of code/space by extending the analysis primarily to how practices of identification sort our physical spaces. Many of the processes covered in the examples below, whether of a distribution center that reads tagged items or a refrigerator that reads cans of Coke, rely on identification. As I show throughout this chapter, without granular infrastructures of identification, many types of animated, networked infrastructure could not exist. As much as discussions of the Internet of Things tend to focus on flashier new technologies, the growth of connected objects first requires the objects to be differentiated. RFID is a primary infrastructure of identification that powers much of that differentiation. Consequently, although the primary focus of this chapter is on the communicative affordances of RFID to connect objects to infrastructure, the underlying argument shows how the expansive growth of identification underlies comprehensive networks of thingly communication.

Identification of "Things" in the Internet of Things

The internet has always been made of "things." The shift with the Internet of Things, however, is in the number and variety of things enrolled in various communicative networks. Following technologist Samuel Greengard and others, if the Internet of Things is defined broadly to include various types of object communication, then one can already find an environment filled with things communicating and connecting with one another. RFID is used in "smart guns" that fire only when in range of a tag,[21] included in casino chips to combat theft,[22] and used in some cars' keyless ignition systems,[23] for just a few RFID-centric examples. Each of these cases involves objects communicating unique identifying information. In the first case, the "smart gun" senses the environment for the presence of the specific RFID tag.[24] Only after that tag's presence is detected will the gun fire. In the final case, the car will start only when it senses the presence of the correct tag. In those cases, the internet is not involved at all, but they include objects communicating with one another to allow a certain behavior.

Most discussions of the Internet of Things focus on communication.[25] But, as argued earlier, underlying that communication is a fundamental process that is less flashy but nonetheless necessary: identification. The layers of connection and communication are undergirded by the process of sorting one object from another. To do so requires the increasingly complex practices of identification covered in chapter 2. Of course, technologies were used for processes of identification long before radar and barcodes, so before moving on to the sorting of connected objects through RFID, I first want to situate that shift within earlier attempts to standardize space through practices of address.

A mostly untold history of the nineteenth and twentieth centuries is the history of addressability. Sociologist Nigel Thrift examined how houses did not have individual addresses throughout most of the nineteenth century.[26] After the creation of individual street addresses, addressability then became more and more fine-grained. The 1960s saw the introduction of five-digit zip codes in the United States, to further segment space. Then, with the adoption of barcodes, the five-digit zip code was transformed into the nine-digit "zip + 4" code to further specialize individual addressability. New practices of address, along with other logistical processes, contributed to what Thrift called the "standardization of space."

The standardization of space refers to quantification and the ability to separate and segment spatial coordinates. That segmentation can occur on fixed representations such as traditional maps, but as Thrift notes, "A second important result has been a change in the nature of the address. Increasingly, addresses are moving with human or nonhuman actants."[27] In other words, the equivalent of unique mailing addresses can now be attached to individual objects through the infrastructures of identification covered in chapter 2. The ability to track and trace the mobility patterns of individual objects contributes to a further standardization of space, one that becomes mobile in ways fixed addresses are not. An RFID system, for example, can collect massive amounts of data about individual locations of objects as they move past readers. The data forms a new type of standardization that focuses on the unique addressability of billions upon billions of bodies and objects as their trajectories are traced through the physical world. Or, as Thrift argues, "Through the standardization of space made possible by these technologies (and the large bureaucracies that employ them), each object and activity taking place on the globe will, at least in principle, be able to be exactly located."[28]

Thrift's concept of the standardization of space was conceived before the widespread popularization Internet of Things. But I want to rethink understandings of the Internet of Things by using the concept of standardization and the technology of RFID to reconceptualize just what an environment of communicative objects *means*. Namely, through infrastructures of identification like RFID, space becomes more standardized as billions of objects become uniquely addressable. Capitalism is defined in part by sameness. One of the fundamental shifts in the embrace of capitalism was the mass production of goods, the ability to make the same thing over and over again. Through RFID, that sameness shifts because each object becomes unique and addressable, even if that uniqueness is found only on numbers on an RFID tag. For space to become standardized at the mass level, objects otherwise awash in sameness must be unique and addressable.

The examples below illustrate the ability to standardize space, to assign unique IDs to billions of objects and record their location as they move through certain passage points. That step back from the surface of connective objects is possibly the major contribution of the larger thrust of this book: namely, RFID as a technology provides a lens through which to view a larger shift in how objects are tracked, traced, and identified and how space is standardized and quantified through those data traces. Later I look at how standardization feeds into data projects that reshape logistical practices, but here I want to stress how the Internet of Things is fundamentally a system of identification even before it becomes a system of communication and connectivity. Each connected object is unique. To understand a voice in an animated network of things, a system must first understand who or what is doing the talking. RFID is an infrastructure of identification that does that communicative work.

The Internet of Things, even as it has been rightly criticized for becoming more marketing buzz than concrete concept,[29] continues to grow quickly. Forecasts predict that thirty billion devices will be connected to the Internet of Things by 2020, though different sources have made predictions that range from twenty billion to fifty billion.[30] Regardless of the exact number, more and more objects will be able to communicate data, and in the discussions that follow, I analyze four broad ways RFID connects objects. First I examine the mobility of human bodies and then look at RFID's role in animating networks of supply, particularly how item-level tagging turns distribution centers and store floors into code/spaces. This discussion expands the definition of the Internet of Things to look at how object communication already shapes our built environment. Then I look at possible future uses of RFID in the Internet of Things by examining two specific areas: the "smart home" vision, which imagines RFID reconfiguring domestic spaces, and the biohacking of subcutaneous RFID injections.

Corporeal Mobility

Cities are made up of individual locations that exist in relation to one another. But cities are also just as reliant on movement as they are on built locations, and the importance of movement has long been neglected in studies of space and place.[31] However, some researchers have begun to move away from more static understandings of space to embrace the role movement plays in how space and place are constructed and experienced. A major part of that move came in what Mimi Sheller and John Urry called the "new mobilities paradigm," which focuses on the importance of movement of all types: people, information, things, and so on.[32] Mobilities research focuses on everything from practices of walking to the mobility of ideas to how the built environment influences movement. As detailed here, RFID technologies already play a role in governing corporeal mobility through their ubiquity in existing mobility systems.

Before looking at how the networking of physical infrastructure influences corporeal mobility, I first want to briefly expand on just why mobility matters. For one, mobility systems are essential pieces of the social construction of space. As geographers Mike Crang and Stephen Graham argue, urban life is shaped by "software sorted geographies" in which software shapes how people move and whether they can move in their desired ways.[33] In some cases, it is software that has contributed to new forms of "splintering urbanism," in which access to infrastructure is differentially distributed, with the fortunate being able to move freely and the less fortunate often relegated to less-desirable forms of mobility.[34] To look at how software and hardware mediate contemporary mobility, the following analysis examines RFID's role in mediating automobility and ridership on public transportation and then extend questions of bodily movement to the role RFID plays in enabling or prohibiting movement in access systems.

Toll Tags and Subway Cards

One of the earliest successful deployments of RFID technology—electronic tolling—is a significant example of "software sorted geographies." The first electronic tolling system was developed in Norway in 1987, followed by the North Turnpike in Dallas, Texas, in 1989. These systems are now ubiquitous and found in countries all over the world, with a few notable examples being the E-ZPass system in the United States, the national eToll system in Ireland, and Singapore's Electronic Road Pricing system, designed to reduce congestion.

RFID-enabled electronic tolling systems include a UHF passive, battery-assisted passive, or active tag attached to the windshield of a car. The tag's unique identifying number is linked to an account, so whenever the tag passes within range of an appropriate reader, the location is logged and the account is charged. The designs of the systems vary, with some tollbooths requiring cars to come to a stop, or at least significantly slow down, while others enable cars to pass by the reader at highway speeds. In essence, every car with an RFID-enabled toll tag can be thought of as an early instantiation of the Internet of Things broadly defined. The cars themselves may not have internet access, but the tag on the windshield communicatively links the car with the larger toll system, serving as a mediator between car, toll infrastructure, and whatever network is used to transmit the collected toll data.

The growth of electronic tolling has had multiple impacts on automobility. For one, as expanded on in chapter 5, RFID toll systems enable new ways to track mobility through data. By assigning each car a unique identification number, toll systems can create comprehensive logs of individual vehicles' mobility patterns. In addition, RFID-enabled toll systems introduce a new actor into networks of automobility. In many cases, the RFID reader and the tag replace what used to be a human actor who collected money or at least was present if something went wrong in a coin system. The toll system remains an obligatory passage point on certain road networks, but now the gatekeeping mediator becomes the automated RFID system.

The automation of these systems does more than raise issues of power in terms of who gets to act in the network by replacing a human with an RFID tag. The systems also are often, if not obligatory, then forced upon the public. For example, in the tollways of my home state of Texas, people do not *have* to participate in RFID tracking. They can pass through tolls and have their license plate automatically read instead. However, they are compelled into these systems because they are charged almost twice as much if they do not use the passive RFID tags for identification; ostensibly, the choice to not participate remains, but the choice is an expensive one.

Automated tolling has also enabled the growth of privatized toll roads that "are only accessible to drivers with in-car electronic transponders."[35] Beyond privatized toll roads, some highways now have dedicated express lanes that are dynamically priced and available only to people with the proper RFID toll pass. For example, the 495 Capital Beltway in Northern Virginia enables certain people to travel on a less-congested road that only accepts E-ZPass payments, and that road is one of many that create a two-tiered form of mobility through the software sorting of movement and identification. Other roads in Northern Virginia, notably a section of Route 66 heading into Washington, DC, are at certain times accessible only to people with

E-ZPass technology; that busy section of road is legally accessible only to people who choose to be sorted by RFID.

The sorting and differentiation of corporeal mobility through privatized and express tollways is part of the larger context of differentiated mobility in contemporary life. Put simply, some people are allowed to move more easily than others through city or suburban streets.[36] More-privileged people drive or take a private car service; the poor often rely on underfunded bus systems. RFID toll tags extend this differentiation to the highways people drive on, and the technology does so because of the combination of software sorting and the major point of RFID's agential effects: the ability to turn almost any object into identification data. RFID toll technologies make cars—and by extension drivers of cars—machine-readable and uniquely identifiable. The machine-readability and unique identification enable the types of software-sorted geographies that shape contemporary spaces and, not coincidentally, enable new networks that sort drivers into tiers based on their willingness or ability to participate in the system.

The sorting that occurs in these "software sorted geographies" raises larger issues about practices of surveillance more generally. RFID tags are technologies of differentiation. They make the mass of vehicles moving at highway speeds identifiable through computing power in the environment. The tracking and tracing of these vehicles through their RFID tags is an explicit form of surveillance that feeds into the standardization of space discussed earlier. The system could not work without the data traces left behind by the RFID technologies. I cover more of the privacy side of toll tags in chapter 6, but an equally important implication of surveillance has more to do with social sorting than with privacy. As surveillance expert David Lyon has argued, social sorting has become one of the major goals of the surveillance society and is enabled in no small part through increasingly complex practices of data collection.[37] The ability to sort cars on the road and to divide in-groups from out-groups based on technology adoption are examples of how practices of surveillance sort populations through identification. These examples also show, through the rather mundane case of toll systems millions of people pass through each day, how the technology of RFID can reconfigure relationships between bodies and larger institutions. A practice that used to be ephemeral—the movement of a car through a coin toll—now leaves permanent identification traces in institutional databases of mobility.

Private automobility is only one type of transportation affected by RFID technology. Public transportation systems of subways, metros, light rails, and buses handle millions of passengers a day. The sheer volume of people who use these systems requires procedures to control, monitor, and properly charge passengers. RFID, most often in the form of short-range, high-frequency (HF) smart cards, has often been the solution. The first RFID public transportation systems were deployed in Asia in the mid-1990s, and RFID is now used in public transportation all over the world.

As mundane as it may be to wave a card to enter a subway, the RFID in these systems enable communicative links between bodies and built infrastructure. They do so because carrying a smart card makes the rider identifiable (assuming they do not lend out their card), in much the same way a toll tag makes a car sensible and identifiable through tolling infrastructure. The typical subway—or bus, in some cases—system works differently than UHF electronic tolling. Smart cards use HF RFID with read ranges of only a few centimeters, so people are not charged just for getting too close to a reader terminal. The short read range, which works similarly to RFID-enabled contactless credit cards, adds a level of human intentionality not present in electronic tolling. Unlike electronic tolling, in which people can pass through at speeds of seventy miles per hour and not realize their movement is being sensed and logged, the short read range of HF cards requires people to actively place the card a few centimeters from the terminal.

Despite their differences from electronic tolling, public transportation systems like the London Oyster card or Japan's Suica cards work through similar means of identification and communication. The built infrastructure wirelessly senses an identifying number on the card, and the number is linked to a personal account from which money is deducted. And here is where these systems relate to the concept of the Internet of Things, broadly defined. Unlike a paper ticket that is physically inserted into a machine, the RFID system identification occurs wirelessly between objects, introducing RFID antennae as mediators between moving bodies and transportation infrastructure.

Smart cards are a common part of the environment for people who regularly ride public transportation, but they might not remain so. Increasingly, engineers are exploring replacing card infrastructure with smartphones that use NFC, a shift that might occur more rapidly now that NFC is available on iPhones. Some cities, such as Barcelona and Singapore, have already begun replacing public transportation cards with smartphones.[38] Regardless of whether smart cards or smartphones work as mediators, however, these systems will continue to rely on some form of contactless RFID to identify bodies and produce data about corporeal mobility. In chapter 5 I look in more detail at the data produced by such systems, but here I want to make the point that RFID technology makes the relatively mundane act of passing through a turnstile an established example of the interlinking object communication of the Internet of Things.

Access and RFID

For about as long as people have moved, bodies have been stopped from accessing certain places. People built walls, castle moats, and of course key locks. All those technologies still exist—well, not many people build moats anymore—and RFID has joined them to become an important technology for controlling access to locations. Many cards, badges, key fobs, and so on contain passive RFID technology that enables entry to certain spaces. For a common example, think of the contactless key cards many hotels use. These cards are programmed with a unique identification that opens a hotel door when the card is sensed by the door's reader. In these cases, the communicative capabilities of RFID tags and readers replace traditional key systems or the contact of magnetic strip cards.

Hotel systems are just one of the many ways RFID technologies are used to either enable or prohibit corporeal mobility. Many offices use RFID badges to control who can or cannot enter. For instance, imagine an employee at a laboratory. She may have a badge she uses to gain access to certain rooms. Each time she waves that badge in front of the reader, the reader authenticates her unique ID and determines whether she is allowed into the room. Her identity becomes readable, and the RFID antenna embedded in the badge acts as mediator between body and reader and database. Unlike a key system, the badge also creates a log of when people enter (identified through the badge's programmed number), raising questions about employee data privacy (examined in chapter 6).[39]

Access badges and contactless hotel keys are commonplace. Many people reading this book have likely waved a card or a badge in front of a door for access. However, that mundanity masks how these relatively ubiquitous RFID technologies represent a shift in physical space and corporeal mobility. An RFID access system embeds hardware and software into the built environment. When someone's card does not authenticate, they are being "software sorted" in a quite literal sense. In addition, if the software or hardware fails, not only are people not able to be automatically sorted, but the space can stop functioning as intended. A hotel room is not much of a lodging if the guest cannot enter; a laboratory is temporarily not a place for scientific work if the doors remain locked. RFID-enabled access systems create code/spaces that will not work without the proper hardware and software. After all, many hotel doors no longer have keyholes as backups; either the reader listens to what the card has to say, or no one enters. The RFID system becomes a vibrant actor in the mundane act of accessing a room.

The access granted or disallowed by RFID systems extends to more than rooms and buildings. RFID technology also plays a major role in the transnational mobility of border crossings. As discussed in more detail in later chapters, ePassports include an RFID chip that contains biometric information used to identify individuals.[40] When someone crosses a border, the agent uses a reader to access the information contained on the chip and match it to the person presenting the passport. The RFID chip becomes a communicative mediator that plays a role in determining whether someone is allowed to move freely across a national border. The border apparatus becomes a code/space when the biometric information contained on a microchip, rather than just the information printed on the informational passport page, is required to allow or disallow entry. RFID is then implicated in larger debates about biometrics—including iris scans, fingerprints, and so on—as their own type of identification system used to sort and classify the population.[41] A tiny piece of mobile technology that remains all but invisible links with the larger border control apparatus and can literally stop someone from entering a country.

As a final point before moving on, I want to look at these examples within the larger context of the Internet of Things. According to some definitions of the Internet of Things, a tag attached to or embedded in an object would not qualify because the object itself does not have computational abilities or an internet connection. But at its core, the Internet of Things is about enabling objects to communicate and building a physical environment that can sense in new ways and react to inputs. Toll tags, subways terminals, access badges, ePassports, and key cards are not new. They are established and ubiquitous at this point. Nonetheless, while they might not be as flashy as a new medical implant or (I guess) a smart refrigerator with a touch-screen display, they are early examples of what the Internet of Things can do in the physical environment. The infrastructure embedded in the physical space senses the presence of an RFID tag and then reacts to that input, whether in the form of billing a toll tag account, pulling back a subway turnstile, opening a locked door, or retrieving biometric information on a custom agent's screen. And these systems all play a role in the increasing standardization of space. Each tag embedded in a plastic card, glued between the pages of a passport, or stuck to a windshield is unique. They differentiate the bodies moving with the tags. They quantify where millions of things are within a larger spatial area at a given moment. The Internet of Things might be most commonly discussed in futuristic terms, but we should not let that futurism blind us to the many forms of object communication that already shape our physical spaces.

The Mobility of Objects

How bodies move is only one part of the mobilities turn discussed earlier; another key piece is the focus on the mobility of objects.[42] The increasingly complicated networks of supply and inventory in the global economy contributed to what I described in chapter 2 as a "control revolution." As James Beniger argued, many technologies of the nineteenth and twentieth centuries arose in response to crises of control.[43] As transportation networks widened and trade became unmanageable, people created techniques for managing and monitoring data flows. These techniques were designed to exert control over the increasingly complicated networks of the Industrial Revolution, and Beniger provided a conceptual framework to look at how innovation often arose in response to shifting mobility networks of supply.

A major technology of control, though one not examined in Beniger's seminal work, was the barcode. Barcodes made objects machine-readable through optical scanning technologies; they were a precursor to today's Internet of Things that linked objects with databases as a form of controlling product data. RFID changes the capabilities enabled through barcodes by networking objects without line-of-sight access and making each object uniquely identifiable. Ultimately, both the identification and communicative abilities of RFID as infrastructure show how object tagging implicates core issues of mobility.

RFID was used to track objects' mobility before Kevin Ashton first presented the concept of the Internet of Things in 1999. However, because RFID tags were expensive, they were used to monitor primarily shipments of high-value products such as nuclear materials and train cars.[44] After the birth of the original Internet of Things concept, RFID began to move to the pallet level to track the mobility of products through the supply chain. For example, Walmart's 2004 mandate discussed in chapter 2 mandated suppliers affix tags to pallets of goods rather than individual goods. In the mid-2000s, RFID technology was still too expensive, and item-level tagging too daunting, for many suppliers to move past the easier task of tagging groups of objects, but pallet-level tagging still implicated issues of mobility and the Internet of Things. A tag affixed to a pallet made that pallet communicative in new ways; the pallet itself had no computational power, but it was able to link wirelessly to computational systems that sensed its presence.

Even as pallet-level tagging struggled to take off in the mid-2000s, some companies experimented with expanding tagging to the item level, especially with higher-cost items. For example, Boeing and Airbus began tagging individual parts for tracking.[45] In addition, a few retailers, such as American Apparel and Marks & Spencer, moved toward a full item-level tagging scheme that made the mobility of each different object trackable throughout the supply chain and on the inventory floor.[46]

So why would a company want to tag objects at the item level? As discussed in more detail in the chapter 5, item-level tagging makes the mobility of the supply chain more visible and granular. Companies are able to track data on individual items through a more automated process than with the optical scanning of barcodes that often required a human. The UHF tags on items moving through a distribution center can be read quickly by readers set up in the physical space. In addition, item-level tagging also has implications related to the networking of objects on store floors. Some of those implications relate to the limitations of barcodes. Barcodes identify classes of objects, not unique objects, and they must be scanned by optical readers.

On a store floor, it takes a significant amount of time to log a complete inventory because people have to scan the barcode of each item. Consequently, many retail stores perform full inventory checks only every few months. That lack of data visibility means inventory accuracy often hovers around 60 percent with barcodes, so companies have only limited knowledge about just where objects are at any given time.[47]

RFID can improve the inventory process through its communicative affordances. Objects affixed with RFID are able to communicate accurately and quickly. For example, someone can scan an individual box of clothes in an incoming shipment in a few seconds.[48] Employees can also walk the store floor and scan entire shelves wirelessly in a short amount of time. The store floor becomes a wireless communicative network, with the tagged items communicating with a handheld reader and the reader feeding data back into a store database. As retail studies have shown, RFID item-level tagging can increase the accuracy of inventory counts to greater than 95 percent.[49]

In addition, because of their longer read ranges and because they do not require direct line of sight, some parts of the inventory process can be automated. One example is the combination of drones and RFID. Warehouses are often huge spaces in which objects can easily get misplace. With barcodes, boxes or products have to be manually scanned and records have to be kept about where the barcode is located in the warehouse. RFID enables semi-automation of this process through drone technology. Small drones equipped with RFID readers can fly through warehouses and inventory tags as they pass by. In 2017, MIT invented a next-generation RFID reading drone that does not even have a stand-alone reader and instead has an RFID chip powered by a remote reader that sends a signal that collects reads from nearby tags. Because the drone has its own RFID chip, its location is able to be triangulated so it can transmit where it is when it picks up reads from nearby tags.[50]

The mobility of retail objects tagged with RFID does more than feed into inventory systems. Some retailers have experimented with novel ways of tracking the movement of objects in retail settings. A controversial example discussed in more detail in chapter 6 is the "smart shelf" technology first piloted in 2002.[51] These shelves included an RFID reader and would take someone's picture whenever the shelf sensed a tagged object moved out of range of the reader. More recently, some high-end stores have begun using "magic mirror" technology that involve RFID readers in dressing rooms.[52] A consumer walks into the dressing room, and the system senses the tagged object and displays product information and related products on the mirror display.[53] The magic mirror shows the potential of the Internet of Things powered through RFID. A tag moving through space is activated by a reader, broadcasts its unique ID, and the system retrieves information displayed on a mirror interface. The object enters into a communicative network and its movement triggers a series of events that link that unique object to the larger data infrastructure of the retail system.

Finally, the tagging of objects not only responds to control crises in retail; RFID is also used to address a different type of control: the real-time tracking of objects in hospitals and construction sites.[54] Hospitals are complex places in which nurses and doctors work across multiple rooms, and pieces of equipment need to be located quickly. One solution to the problem of inventory management has been attaching RFID to equipment ranging from larger machines to technologies as small as scalpels and sponges. These tagged objects broadcast their location to readers strewn throughout the hospital to create what are called real-time locating systems (RTLSs).[55] These systems, also deployed commonly on large factory floors to locate machinery, use RFID to enable objects to communicate their location based on which reader they contact, and that link is visualized on screens to make objects locatable. Quite literally, the mobility of objects becomes networked and tracked, with tags speaking with reader infrastructure to transmit identification data that feeds back into a larger system to display location. In some cases, tracking objects in hospitals extends past locating tools for future use: leaving objects (such as medical sponges) inside bodies is a problem post-surgery. Some hospitals pass a mobile RFID reader over patients to detect if any objects were unintentionally left behind after patients are sewn up.[56]

In these various sites of RFID usage, we can see the consequences of what anthropologist Genevieve Bell and computer scientist Paul Dourish call "yesterday's tomorrows."[57] Bell and Dourish discuss how people focus on idealized technological futures to the point that they do not recognize the parts of those futures that already exist. Much of the discourse surrounding the Internet of Things establishes its own "yesterday's tomorrow," imagining networked domestic spaces, cities that communicate constantly with phones, and so on. But the sites covered in this section are already examples of the Internet of Things, even if they may not be as exciting as some of the future imaginaries covered later in this chapter. A distribution center filled with RFID-enabled objects is a site of constant object communication, an assemblage of things connected to other things that can sort and identify objects on a granular level. Many of these objects do not have their own computing power or IP addresses, but RFID is their link to larger networks where their data can be collected for inventory or their locations can be displayed on interfaces in RTLSs. Factory floors and distribution centers remain out of view for the majority of the population, but they provide a glimpse into just what a future of connected, communicative objects might look like.

The Smart Home

Domestic spaces have always been shaped by technology, whether that technology involves cooking tools, electricity in the home, washing machines, or countless other examples.[58] Much of the hype about "smart homes," however, tends to be rather ahistorical, viewing the home as a site that—until recently—was separate from technological developments.[59] Nonetheless, the development of the supposed smart home extends the relationship between technology and domesticity, which is captured in Frances K. Aldrich's definition of the smart home as

> a residence equipped with computing and information technology which anticipates and responds to the needs of the occupants, working to promote their comfort, convenience, security and entertainment through the management of technology within the home and connections to the world beyond.[60]

Using that definition, and many similar ones, the concept of the smart home is rather broad. On the one hand, some more-experimental smart home projects focused on building environments in which the domestic space reacts to people interactively, sensing their presence and responding by automatically changing temperatures, dimming lights, and so on. But on the more commercial side, much of the smart home discourse focuses on discrete items that often have an internet connection to enable more limited forms of interactivity. For example, my home has a "smart" smoke detector that alerts my phone if smoke is detected. Other examples include internet-enabled "smart" thermostats and "smart" televisions with built-in internet connections.

For all the marketing hype surrounding the smart home, developers are a ways off from the types of interactivity imagined in some of the smart home discourses.[61] After all, a refrigerator with a touchscreen is still a far cry from a completely responsive, interactive domestic environment. At this moment, at least, the concept of the smart home is more like the "yesterday's tomorrows" discussed earlier, partially present in discrete domestic items but still a technological future that remains just off in the horizon, not yet here but positioned as inevitable. RFID has often played a role in the "yesterday's tomorrow" of the smart home vision.

Various experts have identified RFID as a key piece of hardware for the future smart home, for much the same reason RFID is one of the key animating technologies of the Internet of Things more generally: not every object can have its own internet connection.[62] Instead, in some smart home visions, RFID tags (and, increasingly, NFC) communicate with a few readers, and the readers either use the input to make decisions or have their own internet connection and broadcast information to smartphones or other applications.[63] And the positioning of RFID in the smart home is not new. People began writing articles about RFID's domestic potential in the mid-2000s, when it first seemed like item-level tagging was about to take off.[64] Some of those predictions did not age particularly well, but these futuristic imaginaries have now returned, as item-level tagging has finally become a reality in some cases. Here I want to look at a few examples of how the sensing of mobility through RFID feeds into the idea of the interactive home.

One of the most durable imaginaries is that of the smart fridge that has built-in RFID readers.[65] Smart fridge prototypes have been displayed at events like the Consumer Electronics Show (CES), but no models are commercially available.[66] With a smart fridge, individually RFID-tagged items would be monitored by a reader inside the fridge, and when an item is removed, the reader would send an alert notifying whoever does the shopping that an item needs to be replaced. For example, a phone might get an alert when the household runs out of milk or has only two Cokes left in the fridge. Another oft-repeated example is the smart washing machine that can automatically read RFID tags in clothing and adjust washing settings to match the product requirements included in the tag.[67] As a final example, RFID and NFC have also been used to replace keys for homes and adjust home settings based on the individual preferences associated with the ID contained on the tag (e.g., dim the lights or change the thermostat to match the preferences of that specific ID).[68] All of these RFID examples of potential smart home technologies relate back to the Internet of Things and measures of mobility. The fridge sends alerts when it senses an object has traveled outside the read range; the washing machine senses when an object moves within the read range and acts accordingly; the house recognizes the RFID device moving through the front door and alters its settings.

Just like with the other examples covered in this chapter, the smart home examples may seem to raise questions primarily about communication and tracking, but they ultimately represent an imagined shift in practices of identification in domestic spaces. The dream of the RFID-enabled washing machine is that every single piece of clothing becomes uniquely identifiable. The heap of clothes on my closet floor no long becomes a simple mess, but rather a pile of individualized, machine-sortable items of clothing. The smart fridge works the same way. The individual products not only become communicative, they also become identifiable at the item level rather than the class level of barcodes. The domestic space becomes one of finely tuned identification as the relationship among material objects, databases, and humans shifts through small tags attached to objects.

As of early 2018, those types of interactivity and identification are still not found in many contemporary domestic technologies. One reason is that item-level tagging would need to progress far beyond its current state for enough objects to be tagged to make an RFID refrigerator or washing machine viable. But the potential still exists for these types of domestic smart objects, and they remain a topic of research in computer science and engineering. Writing about smart objects, computer scientists Tomás Sánchez Lopez and colleagues created a taxonomy (focusing significantly on RFID) that placed domestic objects in four cascading levels.[69] The lowest level is that of straightforward identification, which is what RFID is often used for. The higher levels focus on "decision making" and "networking"; in other words, higher-level domestic smart objects will be able to network among devices and use software to make decisions based on the data they receive. The case of the washing machine, in particular, showcases a "higher-level" smart object than most contemporary smart home objects that focus more on alerting people than on making a decision. The washing machine would be able to use the RFID data to automatically run processes set to wash clothes, removing human decision making from the domestic task.

Of course, just because a higher-level domestic smart object is technically possible does not mean it is necessarily useful. One of the problems with smart home technologies more generally has been that the people who design the technologies tend to be male and the people who disproportionately perform domestic tasks still tend to be female: "Women have long been disenfranchised from the development of the domestic technology they use, playing little or no part in the design process which generally views them as passive consumers."[70] Or, to put it differently, the smart home is a "gendered sociotechnical construction developed in line with the interests of its male designers."[71] These criticisms of the gendered nature of domestic technology could possibly explain why the adoption of smart home technology has continued to lag. Even if item-level tagging did progress far beyond its current state, it is unclear whether the people who typically do the grocery shopping or wash the clothes even want these systems.

As a final point, RFID-enabled smart objects are almost always presented in terms of their convenience as time-saving domestic technologies. But, if RFID ever does take its place in the assemblages of domestic life, there is reason to be skeptical about how technologies like smart fridges or washing machines affect domestic labor. As historian of technology Ruth Schwarz Cowan detailed, many domestic technologies, ranging from stoves to new distribution techniques for food to washing machines, were marketed as easing domestic labor.[72] Instead, however, many of these technologies redistributed domestic labor, lessening it for men and children while increasing it for the female heads of household. For instance, washing machines raised expectations of cleanliness that fell upon women and often increased their labor in spite of technological advances. If RFID-enabled technologies do enter the networks of domestic life, researchers need to be attentive to the gendered impacts of domestic technology and look critically at how higher-level smart home objects affect practices of domestic labor.

Biohacking

Many of the RFID technologies discussed in this chapter are found inside objects. Smart cards have antennae lodged between plastic; packages can have RFID inserted between layers of cardboard; smartphones have NFC built in. The embedded nature of RFID is part of what makes the technology fairly ubiquitous yet relatively invisible. Possibly in no case is RFID more invisible than when it is embedded in a body. RFID has been injected into pets and livestock for more than two decades,[73] but the technology has also made its way into human bodies within biohacking communities.[74] If an article refers to a "microchip" implant, chances are the microchip is some form of RFID.

Subcutaneous RFID is an example of the Internet of Things brought to a transhumanist level, merging the human body with communicative technology that enables connections to the physical environment. These forms of bio (or body) hacking represent the push toward the cyborg, which science and technology studies researcher Donna Haraway defined as a "cybernetic organism, a hybrid of machine and organism, a creature of social reality as well as a creature of fiction."[75] And RFID implants have an established history as part of the cyborg imaginary: in 1998, Dr. Kevin Warwick became the first human to inject an RFID chip, which he later declared made him history's first cyborg.[76] Artists such as Nancy Nisbet and Lukas Zpira also implanted RFID to explore issues of surveillance and transhumanism.[77]

Before delving more into biohacking and subcutaneous RFID, I want to note some of the ahistorical and gendered nature of discussions about microchip implants. The "cyborgs" with microchips are often portrayed as revolutionary examples of using internal technologies to alter bodies' relationships to the world. However, some of the coverage of biohacking seems to forget that the body/technology relationship goes back much further than the 1990s. People have used pacemakers to regulate heart rhythms for many decades, and in the mid-twentieth century, women began implanting intrauterine devices (IUDs) that govern hormones and thicken the mucus wall of the cervix. Implanting a device that literally controls one's hormones and prevents procreation might be more of a technological marvel than an RFID chip someone can use to open doors or make credit card payments.

That being said, the use of RFID inside the body shows a melding of computing and bodies that gained some attention even outside niche communities in the mid-2000s. In 2004, Mexico's attorney general and 160 of his employees injected RFID that worked as access control for secured rooms.[78] That same year, the story of a bar in Barcelona that let people inject RFID and use their chips to pay their tab received international attention.[79] A company called VeriChip (now rebranded as PositiveID) received FDA approval in 2004 for their implantable RFID chip. A 2006 *Daily Mail* article even included the warning headline, "Britons 'Could Be Microchipped Like Dogs in the Next Decade.'"[80]

Ten years later, I am not going out on a limb to assume that few people reading this book have a subcutaneous microchip. However, injectable RFID has found its place in certain biohacker communities. One of the most public proponents of subcutaneous RFID is Amal Graafstra, a self-proclaimed "double RFID implantee."[81] Graafstra has written books on DIY RFID projects and has a series of videos about using RFID implants to interact with the environment.[82] The videos show an embodied form of the Internet of Things in which the wave of a hand links the body to a reader and connects to a computer or unlocks a door. Graafstra's online store, Dangerous Things, sells "custom gadgetry for the discerning biohacker," including RFID transponders and medical equipment for injections.[83] Other biohackers have also experimented with RFID injections to network bodies with infrastructure. For instance, the Swedish biohacking group BioNyfiken provides employees at an office park with optional RFID injections.[84] The injected chip becomes the means of identification to connect with the office park's infrastructure and enable access control and identification to use photocopiers and so forth.

Outside biohacking communities, RFID implants in humans are rare, though they have been used in a few controversial medical cases. Possibly the most controversial was a 2007 case in which VeriChip injected two hundred Alzheimer patients with RFID without receiving proper authorization. Privacy groups protested when they found out about the pilot project, and VeriChip eventually changed its name to PositiveID, in part because of the controversy.[85] Although researchers do still discuss using RFID in medical situations, they now focus mostly on wearable RFID bracelets rather than injections.[86] Outside the medical field, one of Sweden's rail lines announced in 2017 that they will start accepting payments from subcutaneous RFID.[87] It is not clear how many people took them up on the opportunity, though the number likely falls in the low thousands.

Stories of injectable RFID still get outsize attention whenever a new case arises, however, despite the relatively niche status of the practice. As I wrote an early draft this chapter, stories were being shared all over social media about a company called Three Square Market that offers employees optional

RFID implants. The chip, which is about the size of a grain of rice, enables employees "to make purchases in their break room micro market, open doors, login to computers, use the copy machine."[88] Despite the fact that the implant is optional and is expected to affect only around fifty employees who choose to undergo the relatively simple procedure, the story was picked up by many major news outlets in the United States, including the *New York Times, Washington Post,* and *USA Today.* That case is worth mentioning here because it is a reminder of the high level of interest—and distrust—people have in regard to bodily implants. One small company can make national news through a voluntary implant program that will affect fewer than one hundred people.

The use of RFID as implantable technology for humans remains rare. However, despite the limited spread of injected RFID, the technology has become a major focus for some more conspiracy-minded websites. Here I want to spend some time focusing on how certain communities have turned the identification capabilities of RFID technology into a synecdoche for the surveillance state in general. I argue that the fears about RFID, particularly injectable RFID, capture the larger discourses surrounding communicative objects and infrastructures of identification. Many of the conspiracy theories are false or do not understand how RFID works. In particular, some of the conspiracy theories confuse RFID with GPS despite the fact that passive RFID has no "tracking" capabilities itself unless the tag is read by a reader. But, as I discuss below, in some of these conspiracies one can find larger concerns about how processes of identification reconfigure relationships among bodies and institutions.

Most of the conspiracy theories about RFID focus on subcutaneous RFID chips and, unsurprisingly considering many of the sources, talk about government agencies' ability to read bodies and assign them unique IDs for tracking purposes (kind of like a toll tag injected into your bicep). Alex Jones—likely the most influential conspiracy theorist in the United States—has been a particular critic of imaginary RFID plans. At various points, Jones's websites, InfoWars and Prison Planet,[89] have claimed that California had begun implanting homeless people with RFID,[90] that the European Union and United States governments will soon begin mandatory RFID implantation,[91] that the Swedish government has announced plans to microchip the entire population (and abolish genders),[92] that RFID was being used in twenty-dollar bills to track the population and could explode if microwaved,[93] and that RFID chips were being hidden inside breast implants so people could be tracked.[94]

Jones's various sites are maybe the most prominent in espousing theories of wide-ranging surreptitious RFID implant, but they are far from the only ones. One fact-checking site alone, Snopes .com, features at least ten refutations of various RFID-implant conspiracy theories that have spread across the internet. Here are just a few the site has debunked:

- "Australia Becomes First Country to Begin Microchipping Its Public"[95]
- "RFID Chip Implemented in All Public Schools by 2015—Effort to Curb Gun Violence"[96]
- "Study Finds 1 in 3 Americans Have Been Implanted with RFID Chips: Most Unaware"[97]

The conspiracies about RFID fit within existing narratives about the creeping surveillance state. RFID is used to further these narratives, and the fact the chips are small and can be injected enables sites to fan the flames of fear about identification technologies. One final RFID conspiracy theory shows how fears about the identification capabilities of subcutaneous chips can be utilized for political purposes: the "Obamacare" chip. In the United States, 2009 was marked by discussions of "Obamacare," the at-first-derogatory and now embraced nickname for what was eventually officially called the Affordable Care Act (ACA). Conservative political activists organized to protest potential health-care reform, in no small part powering—along with early-2009 bailouts of banks—the growth of the far-right Tea Party movement in American politics.

RFID tags would not seem to have much to do with debates about health-care reform. However, through a strange series of events, RFID played a significant role in Obamacare conspiracies in some far-right communities. The root of the larger conspiracy was a short clip from a 2007 NBC News segment called "Life in the U.S. in 10 years time."[98] The first prediction of the segment featured a voiceover saying,

> The year is 2017. You're rushed to a hospital with no ID and no medical history, but thanks to a microchip under your skin, it's all there. Science fiction 20 years ago, but a biometric reality today.

The video clip included footage of a doctor using an RFID reader on a human arm, and that first scene was followed by an expert who describes the benefits of using microchips to identify people. The clip is less than three minutes long, but it was then repurposed with YouTube titles like "NBC ANNOUNCES EVERYONE MUST WEAR A MICROCHIP BY 2017."[99] The clip was also referenced again and again in various articles about RFID implantation and the ACA, a conspiracy theory that then grew because of a misinterpretation of an early draft of the bill. That early draft was called HR 3200 and did not pass the House of Representatives. As seen in the chain email included below, HR 3200 did have language about collecting data about medical devices "used in or on a patient," but that data was to be used to track device efficacy and had nothing to do with mandatory RFID implantation. Regardless, that language never made it into the bill that passed Congress, so the point is moot. Nonetheless, there are many articles on far-right sites and videos on YouTube about how Obamacare requires mandatory RFID implantation that will be used to identify every member of the US public. The fears culminated in a chain email that went viral and had to be debunked on multiple fact-checking sites and news channels.[100] The chain email is included in full below:

> Seems people will be getting more than they bargained for in this new health care bill.
>
> The Obama health care bill under Sec. 2521, page 1,000 will establish a National Medical Device Registry. What does a National Medical Device Registry mean?
>
> National Medical Device Registry from H.R. 3200 [Healthcare Bill], pages 1,001–1,008:
>
> (g)(1) The Secretary shall establish a national medical device registry (in this subsection referred to as the 'registry') to facilitate analysis of postmarket safety and outcomes data on each device that:
>
> (A) is or has been used in or on a patient;
>
> (B) and is: (i) a class III device;
>
> or (ii) a class II device that is implantable, life-supporting, or life-sustaining.
>
> A "class II device that is implantable?"
>
> Then, on page 1,004, it describes what the term "data" means in paragraph 1, section B:
>
> "(B) In this paragraph, the term 'data' refers to in formation respecting a device described in paragraph (1),including claims data, patient survey data, standardized analytic files that allow for the pooling and analysis of data from disparate data environments,

> electronic health records, and any other data deemed appropriate by the Secretary."
>
> Approved by the FDA, a class II implantable device is a "implantable radiofrequency transponder system for patient identification and health information."
>
> This sort of device would be implanted in the majority of people who opt to become covered by the public health care option. With the reform of the private insurance companies, who charge outrageous rates, many people will switch their coverage to a more affordable insurance plan. This means the number of people who choose the public option will increase. This also means the number of people chipped will be plentiful as well. The adults who choose to have a chip implanted are the lucky (yes, lucky if you're a Govt Control Libtard) ones in this case. Children who are "born in the United States who at the time of birth is not otherwise covered under acceptable coverage" will be qualified and placed into the CHIP or Children's Health Insurance Program (what a convenient name). With a name like CHIP it would seem consistent to have the chip implanted into a child. Children conceived by parents who are already covered under the public option will more than likely be implanted with a chip by the consent of the parent. Eventually everyone will be implanted with a chip. And with the price and coverage of the public option being so competitive with the private companies, the private company may not survive.

The chain email was sent to thousands of people, and at one point, the *RFID Journal* even ran an article debunking claims that Obamacare requires mandatory implantation.[101]

On the one hand, the conspiracies about mass implantation can be dismissed as worries from fringe groups. But, on the other hand, a few things about that: for one, people like Alex Jones have weekly listenership in the multimillions, so they are not as "fringe" as many people would like to think. For another, while I have mostly looked at fringe sites, the idea of implantation has slowly but surely gained increased attention outside these communities. For example, a 2017 article in *USA Today* ran the headline "You Will Get Chipped—Eventually." The article featured a quote from University of Wisconsin associate professor Noelle Chesley, claiming, "It will happen to everybody.... But not this year, and not in 2018. Maybe not my generation, but certainly that of my kids."[102] I have my doubts about that argument, but it is certainly possible.

Chapter 6 covers issues of privacy and surveillance, so the question remains why I covered these conspiracy theories about RFID implantation here. I did so because the issue of bodily implantation captures on a deep level the dual issues of communication and identification. Through a subcutaneous chip, human bodies can be linked directly to infrastructure just like any other form of (mostly) HF RFID. Whether biohackers or conspiracy theorists, their hopes or their fears are centered on a convergence through RFID: a move to reshape the flesh-and-blood body into the center of various media ecospheres. The Swedish subway riders turn their palms into subway cards and some biohackers transform their flesh into a credit card or their skin into a key. The conspiracy theories about RFID fear convergence as well: Our bodies will become walking identification numbers. Each body will be uniquely identifiable through machines and transcribed with codes that grant access to everything from police records to our medical records, in the case of the "Obamacare chip." Just as so many media forms converge through smartphone technology, the ambitions of biohackers and fears of conspiracy theorists focus on many currently discrete forms of information access converging in a tiny chip buried under the skin.

Widespread RFID implants are likely still a long way off. But in the fears and potentialities of these niche communities we can see a distillation of the core arguments shaping this book: namely,

RFID can simultaneously animate the environment through novel forms of object communication and turn almost any material thing (including bodies) into identifiable, trackable data. And while these communities focus on injectable RFID, much of their larger discussion can be applied to other forms of RFID, ranging from wristbands to cards people carry in their pockets. The unique thing about subcutaneous RFID is its supposed permanence, but anyone with a semi-strong stomach can look up videos of people removing subcutaneous chips with a scalpel.[103] Consequently, I argue that in both the embrace and the fear of microchips one can find a larger discussion about the consequences of identification writ large: these chips leave a semipermanent, communicative identification trace on the body, one that can be read by the built environment.

I do not know whether widespread implantation will become common anytime soon. If it does happen, it will hopefully happen with limits on the data practices of institutions, though I am not optimistic that would be the case. But it will also be important to understand just how these microchips work as part of the Internet of Things. HF RFID has short read ranges. It is not a long-range technology like UHF RFID, so the tracking potential is different. However, I do not want to dismiss the idea of subcutaneous RFID as a future piece of the Internet of Things. I return to the topic of implants in chapters 5 and 6, but I want to end here by noting that the Internet of Things *could* move inside our bodies, whether as a way to identify people or to allow people to interact with software with no mediator other than an RFID chip and a layer of skin. Biohackers have already shown it is possible, and the small, often hidden nature of RFID technology shows why these chips have long played a significant part in how some communities understand the future of the Internet of Things and transhumanism.

Identification in the Internet of Things

Far more things than ever before can communicate data, and the number will continue to grow in the near future. Those two points are fairly inarguable. What an increasingly connected environment will look like and what the consequences will be, however, still remains in doubt. One of the arguments of this chapter has been that, to help understand the possible future of the Internet of Things, people can look at the various forms of object communication that already exist. After all, physical environments already contain objects that communicate with each other and with infrastructure. Some objects do so through their own internet connection, and the section on IPv6 in chapter 2 examined how IP addresses work as an infrastructure of identification. Many other objects do not and will not have their own internet connections and instead communicate through RFID. Even if smartphones replace subway cards or contactless credit cards, the NFC they use to communicate will be a type of RFID. The past, present, and future of at least some parts of the Internet of Things involves RFID.

The forms of RFID covered in this chapter are more technologically limited than many internet-connected objects. People cannot use RFID to download firmware updates, connect objects with smartphones from miles away, log in to home cameras from work, or change settings through a website or a mobile application. Some technologies will combine internet connections with RFID, such as the "smart fridge" discussed above that would use RFID for identification and the internet to send alerts. But as the examples in this chapter have shown, many forms of object communication will not initially involve the internet, instead relying on RFID to connect and identify billions of objects. After all, many networked objects are too plentiful, too small, or too cheap to get their own internet connection. The idea of a ten-cent printed tag uniquely identifying an object might not seem transformative on its own, but when those forms of identification extend to the tens of billions, it shows the vast potential for object communication in the Internet of Things.

I began this chapter by looking at different definitions of the Internet of Things and adopting Weber and Wong's argument that "the 'things' that Internet of Things will connect subsume and go beyond devices with computational capabilities."[104] Building on that point, the above analysis showed how objects that themselves often do not have computing power use RFID to link with larger computational systems, whether in the form of mobility, logistical, or retail infrastructure. The objects transmit identification data to infrastructure built into the environment that can then react to inputs. Importantly, much of the data transmission with RFID can occur with minimal human involvement, and that sometimes minimal involvement is one of the major shifts from barcodes to RFID.[105] With barcodes, humans have to optically scan an object. They are active participants. With RFID, particularly in the supply chain, objects can transmit data without direct human involvement. Even RFID technologies such as smart cards or access badges require less conscious effort than technologies like magnetic strip cards or physical keys.

The out-of-the-way nature of RFID is part of what makes the technology so interesting. Before I started researching RFID, I never stopped to think about how I got through a subway terminal, drove through a toll without stopping, or made a contactless payment. Most people I talk to about this project do not know what RFID is, even though many of them use some variation of the technology nearly every day. Consequently, the RFID systems discussed above fit well with what computer scientists Mark Weiser and John Seely Brown called "calm technology,"[106] the goal of which was for people to "simply use them unconsciously to accomplish everyday tasks."[107] Calm technology developed out of Weiser's broader writings in the late 1980s and early 1990s about ubiquitous computing (ubicomp) that envisioned a future in which computing would be built into the environment and fade into the background of people's interactions with physical space.

RFID systems, at least when designed well, often work as a kind of calm technology that makes things happen without requiring too much attention. Weiser and Brown's three broad guidelines for designing calm technologies can illustrate why:

1. A calm technology should reside mostly in the periphery of the user's attention but should be able to quickly move briefly to the center of attention.
2. A calm technology should increase people's "peripheral reach" by improving what they can do in the environment, but without burdening them with information.
3. The technology should become familiar and comfortable and not distract from one's surroundings: "The result of calm technology is to put us at home, in a familiar place. When our periphery is functioning well we are tuned into what is happening around us, and so also to what is going to happen, and what has just happened."[108]

Even after thirty years, Weiser's visions of ubicomp and calm technology continue to shape developments in computer science.

Entire conferences and journals focus on ubicomp, and the concept of calm technology still influences many visions of what the Internet of Things will look like.[109] However, Weiser's writings on ubicomp are more than a prediction that some people inevitably think will come true; his writings about technology fading into the background have become a sort of self-fulfilling prophecy that shape development just as much as they predict it. In addition, as Paul Dourish and Genevieve Bell argue, the vision of a seamless, ubicomp future has blinded some experts to the fact that computing is already ubiquitous.[110] The example the two authors focus on is mostly the mobile phone, and they argue that mobile phones are computers and already everywhere. Smartphones, in fact, are more advanced than many of the forms of physical computing Weiser predicted.

Many of the RFID cases covered in this chapter also fit with Dourish and Bell's arguments about how visions of the future of ubicomp obfuscate the fact that the future, in many ways, is already here. Many pieces of the physical environment are dependent on computing, even if the computing is less informational and end-user-focused than some visions of ubicomp. The corporeal mobility discussion in this chapter in particular made that case. In many parts of the world, RFID systems are everywhere, and the computing in those systems is a key piece of various infrastructures of mobility. RFID systems open subway gates, process toll traffic, and enable access to buildings and rooms. The technology also makes computing more ubiquitous in the supply chain by connecting billions of objects, which may not have any computing power themselves, to larger systems. As argued earlier in this chapter, discussions of the Internet of Things or ubicomp more generally may often focus on futuristic technologies, but that gaze into the future sometimes ignores the already ubiquitous forms of computing that shape people's lives.

In addition, the role RFID plays in the Internet of Things harkens back to Weiser and Brown's discussion of calm technology, possibly more so than smartphones and other forms of mobile computing. Smartphones may be everywhere, but they are not "calm." They require attention and have been criticized for being too often at the center rather than the periphery.[111] RFID technologies, on the other hand, often quickly move from the periphery to the center for a brief moment. Take an access control card as an example. Someone needs to enter a door, they momentarily take out the card, and then the card goes back in their pocket. In Heideggerian terms, the RFID technology is ready-to-hand and requires little attention or engagement about just what the larger RFID system *is* or how it works. People wave an RFID card of some type and something happens. Little extraneous information is presented through the system.

Of course, although a smart card might require little attention from a user, the technology does not fade completely into the background. Writing about the technologies in ubicomp, Weiser argued that "a good tool is an invisible tool. By invisible, I mean that the tool does not intrude on your consciousness; you focus on the task, not the tool."[112] A few of the RFID technologies covered in this chapter do possibly rise to the level of invisibility. Take the toll tag as an example. Someone who has an RFID toll tag on their car and drives through a toll at highway speeds might not even realize their tag is being read. They can focus on the task of driving and do not have to alter their behaviors to interact with the larger system. The same is true of UHF tags attached to pallets or items that may be scanned automatically as they move through a loading door equipped with readers: if set up well, the connection occurs seamlessly. Other examples, however, are not as seamless. When someone waves a card in front of a payment terminal or subway card, or even waves their palm, in the case of biohackers, the system does demand at least some of their attention on the tool. That at least minimal attention, however, will possibly be a key piece of the ethical development of the Internet of Things.

Some views of ubicomp and some predictions about the future Internet of Things focus on invisibility and seamlessness,[113] while other computer scientists argue that ethical development will require built-in "seams." Weiser himself, who was aware of the potential privacy and surveillance implications of a world of connected objects, argued so. Another was Matthew Chalmers,[114] who drew from Weiser's work to argue that ubicomp should incorporate "seamful design," in which people are made aware of different computing systems. Building on his argument, I want to conclude here by arguing that the "seams" visible when someone has to wave an RFID technology in front of a reader are important for the design of the Internet of Things. That simple action at least makes people aware they are interacting with identification infrastructure, in contrast to systems like toll tags that let people drive through a toll without any acknowledgment or alert. As discussed in the biohacking section and in more detail in chapter 6, many of the fears about the identification capabilities of RFID focus on seamlessness and invisibility. People are afraid that RFID technologies, whether implanted in the body or not, will be read surreptitiously without their knowledge.

The idea of RFID as seamful design harkens back to the argument in chapter 3 about the necessity of understanding the basic difference between types of RFID. A subway terminal is seamful because it uses HF technology that has a short read range. A card with a UHF tag could conceivably enable people to walk through a terminal without taking out a card or performing any action because the card could be read from a distance. At different points, seamlessness has been the goal of some more lofty RFID predictions, such as the goal of a seamless shopping experience, in which people put UHF-tagged products in their shopping cart and then stroll past readers that automatically inventory their purchases and debit their account.[115] But for as much as invisibility might be convenient in the Internet of Things, it also raises major issues of surveillance that more "intentional" technologies, such as an implanted HF RFID chip, do not. And the possible sacrifice of seamlessness for intentionality is not a bad thing. In a distribution center, a design that automatically identifies tags when they move past an embedded reader can be useful. In cases of corporeal mobility, however, invisibility is not necessarily something to strive for. The data on RFID tags can be linked back to people, so systems should make people aware when the identification data is being transmitted. As long as mobility systems (except for toll systems) continue to use HF RFID—including NFC—some intentionality will be necessary. However, if future mobility infrastructures move to UHF technologies, people should be made aware of how the shift affects practices of identification.

Conclusion

The Internet of Things will involve internet-enabled devices, but it will also include billions of objects made uniquely identifiable through technologies like RFID. In the near future, some of the examples covered above will likely be replaced by the NFC in smartphones now that Apple has made NFC more available, but the technology of identification will still be a type of RFID. Even if NFC does become more popular, RFID's role in the Internet of Things will continue to be less obvious and less flashy than the growing number of internet-connected objects. However, as I have shown in the examples above, part of what makes RFID unique is how calm the technology can be as it connects objects and bodies to infrastructure and sorts and categorizes difference in the physical world. The increasing standardization of space requires specific, detailed logs of the mobility of billions of people and things, and as covered in chapter 5, RFID provides a prominent voice that shares data about everything from the movement of livestock to the authenticity of pharmaceuticals.

5 Data Traces of Identification

Many of the things I do every day leave little trace on the world. No lasting record exists of the coffee I just poured, the door I just opened to let my dogs out, or the eggs I just made. Some of my other activities, however, do leave behind traces of data. The websites I visit are logged by my ISP and the cookies in my browser, my location data is transmitted to my cellular provider multiple times per minute, and my word processing program creates metadata about this chapter as I type. I leave behind digital data traces every time I use my laptop or my mobile phone to do pretty much anything. But I know all that. What I had not considered until I began researching RFID was how many of my direct interactions with physical infrastructure left behind data as well.

I want to recount two weeks of my life in mid-March 2017 to give a few examples. During that period, I took a trip to Japan. When I was in Japan, I used a Suica card to ride the Tokyo metro. The card uses RFID and leaves behind a data trace whenever I enter a gate. I ate some conveyor belt sushi that possibly used RFID tags on plates to track data that shows how long the sushi remained on the conveyor belt.[1] When I got back to the United States, I passed through border control using the biometric data contained on the RFID chip in my passport. I retrieved my car and left the parking lot using my RFID-enabled toll tag. I stopped to get food on my way home and used an NFC app on my phone to pay. Each of these interactions involved data transmitted wirelessly through RFID technologies.

The examples above are just a few instances of how RFID data is used to track various "things" in the physical world. Other uses of RFID involve data on much larger scales. After all, RFID technologies produce massive amounts of data. Whenever a passive tag is within range of a reader, it transmits data and keeps transmitting data until it leaves the read range. The data might consist of a serial number used to access account information in a database, or it might include a full EPC that includes a company code, product code, and unique identification number. The data might be far more extensive and involve sensors that can transmit everything from the chemical levels in food to the temperature levels in the Urine Processor Assembly on the International Space Station.[2] To put the scope in perspective, a large system with around a thousand readers and ten thousand tags can generate more than a terabyte of data daily.[3]

This chapter focuses on RFID data, particularly big data, a concept I explain in more detail in the discussion that follows. The sheer amount of data people produce has reached levels that are difficult to conceptualize. For example, "Google processes more than 24 petabytes of data per day, a volume that is thousands of times the quantity of all printed material in the US Library of Congress."[4] Experts have labeled our current moment the "petabyte age" to give perspective to the massive growth in data.[5] But even the idea of a "petabyte age" quickly became outdated, with people arguing we are already entering the "zettabyte age."[6] RFID plays a role in moving us to that "zettabyte age," turning things ranging from bodies in Disney World to pharmaceuticals in the supply chain into pieces of larger data infrastructures. But big data is not the sole focus of this chapter. A few of the examples I cover are smaller, more individualized interpretations of data, particularly individual fears in evangelical Christian communities about RFID.

Broader discussions of data raise questions that are often elided. Namely, just what is data and what can data do? And what is so different about big data? Consequently, I next look at data in detail before transitioning to the epistemological shifts heralded by the supposed "big data revolution." After discussing big data, I provide examples of seven different areas—ranging from RFID's connections to blockchain technology to fears about the "Mark of the Beast"—in which RFID data is used to produce various forms of data visibility. By no means is the analysis meant to be a comprehensive account of all the ways RFID is used to produce identification data. Instead, I go into some depth on a few cases to capture the wide scope of RFID as data-producing infrastructure. The goal of this chapter, however, is to do more than just talk about RFID in terms of data. Instead, I want to use RFID as a lens through which to view the increasingly important role data plays in the contemporary world, while also pushing back against some of the rhetoric about data "speaking for itself."

The Meaning of Data

Data is a word, like "privacy" or "community," that is used frequently but rarely defined. Data has different meanings and definitions depending on the discipline, but at its core data is a base used to build arguments and progress toward higher levels of understanding. As David McCandless put it, at the base of the knowledge pyramid is the known world. Data is the next level up, representing abstractions of pieces of the world. Those abstractions are collected in discrete form and then combined to form information, which then ideally leads to knowledge and then wisdom: the application of knowledge. Without data, the knowledge pyramid is never built because the world is never abstracted in an analyzable way.[7]

Data is often portrayed as a neutral representation of the world, as a value-free abstraction of a phenomenon. But data "are more productively understood as both a component of the real and a producer of the real."[8] In other words, data is not just a representation of how people understand reality; it also shapes our understanding. In addition, data is never "raw" in the sense of being objective and free of human influence. Instead, "raw data is an oxymoron,"[9] and data can never be conceptually removed from the technological, social, economic, and ethical frameworks in which it is collected. In a detailed account of how data collection is shaped, philosopher Bruno Latour's book *Science in Action* documents the many practices of interpretation and technologies of collection and analysis that influence how scientific data is understood.[10] Many other social scientific and humanistic analyses have made similar arguments, showing how data is a human construct that cannot be separated from processes of interpretation. To be clear, I am not arguing for an extreme view that no reality exists free of human interpretation; instead, I am arguing that people cannot understand reality without layers of both conscious and technological mediation. There is "much more to conceptualizing data than science and business generally acknowledge."[11]

Data has been an important part of discovery and knowledge for centuries. Recently, however, data has gained increased attention with the popularization of the "big data revolution,"[12] a movement that has both led to increased RFID adoption and been partially shaped by RFID technologies. However, before moving on to a discussion of the tenets of the supposed big data revolution, I first want to provide brief historical context: namely, the "big" in big data is relative. Data has often seemed big. People complained about the "bigness" of printed data within a century of Gutenberg's printing press;[13] communities developed complex accounting and management techniques to manage data in response to control crises that followed the Industrial Revolution;[14] and nations developed census techniques to rationalize and monitor population growth as populations grew within national borders.

Despite some of the ahistoricism of big data discussions, the use of the term "big data" does capture a shift in how pervasive data is in contemporary life. Two related developments are primarily responsible for that shift. The first is the growth of computing power. Many of the earliest computers were built as calculators to handle data, but they did not have enough processing power to handle massive amounts of data or make complex calculations. Computing has come a long way over the last sixty years or so, but even now, some particularly large data projects require special computers to handle data analysis. The second development is the sheer amount of data produced every day. Every time someone goes to a webpage, drives through a toll, sends a Facebook message, and so on, their actions produce trackable, analyzable data. RFID technologies are part of that trend, producing far more data than traditional barcode systems.

Big data is most often talked about in terms of volume, with people focusing on the massive amounts of data collected in various projects. But volume is only one difference in contemporary big data practices. After all, while data sets have grown larger, huge data sets existed long before the 2000s in the form of national censuses and large-scale scientific collaborations. Consequently, the shifts with big data focus on what are often referred to as the three Vs: volume, velocity, and variety.[15] With a national census, the scale of the data collection might be huge, but because the data is collected only every ten years, it does not have the velocity of contemporary projects that are able to collect data in near real time. RFID is an enabling technology of big data projects because it enhances each of these three Vs. The data produced through the technology can be huge in scope (volume), can be collected more rapidly than optically read technologies such as barcodes (velocity), and can contain more types of data than earlier identification technologies because of the larger storage capacity of RFID tags (variety).[16]

Identifying the three Vs of big data still does little to explain just what big data represents. Proponents argue that big data, in its more "revolutionary" forms, is not just about increased data visibility; it instead represents an epistemological shift in how we know the world.[17] That idea was popularized in a 2008 *Wired* article by Chris Anderson with the rather dramatic title, "The End of Theory: The Data Deluge Makes the Scientific Method Obsolete." Anderson argued that, with the growth of big data, people would no longer need the scientific methods of observation, hypothesis, and analysis. Instead, researchers can do away with models and instead look for correlations in huge data sets. It will no longer matter *why* something happens, only that it happens.

Anderson's arguments were expanded upon and tempered in a book by Viktor Mayer-Schönberger and Kenneth Cukier called *The Big Data Revolution*. The book provides a detailed discussion of the epistemological shifts of big data, and the authors argued that with "much more comprehensive data sets we can shed some of the rigid exactitude in a big-data world."[18] The authors also argued that "society will need to shed some of its obsession for causality in exchange for simple correlations: not knowing why but only what."[19] Key to both these works and others about big data is the belief that with large enough data sets, "data can speak for themselves."[20] Once data becomes large enough, it supposedly rises to a new level of objectivity and can be analyzed and acted on free of human bias and interpretation. However, the idea that "data can speak for itself" has been criticized in humanistic and social scientific circles.[21] Consequently, I want to conclude this section by looking at two criticisms I return to when I look at RFID in the context of the push toward big data (I wait until chapter 6 to cover privacy issues): (1) proponents sometimes do not acknowledge the limitations of data, and (2) data always requires interpretation.

The first criticism concerns the idea of total visibility and objectivity of data. In an ideal world (at least, ideal for some data scientists), data would provide an entire picture of a phenomenon, whether that phenomenon is something like urban mobility or the history of a pair of jeans in a global supply chain. However, data is not produced free of human involvement. The choice of what data to collect—for example, where to deploy RFID readers or what information to associate with a serial number on an RFID tag—is intentional and limiting. Rarely could anyone collect all data about a phenomenon. Based on technological limitations, budgets, political climates, regulatory frameworks, and so on, almost all data sets—no matter how comprehensive—leave something out. External factors often shape what is collected, whether those external factors spur or hinder new types of data collection. Throughout the discussions that follow, I return to how regulations and economics influence RFID deployments to show why discussions of "data speaking for itself" ignore the host of factors that influence how data is produced. In sum, the "total visibility" of the technological sublime of RFID is almost never possible, which is not a problem in itself but is a problem when potentials are overstated.

The second issue with some of the discourses surrounding big data is the lack of recognition that all data needs to be interpreted. One of the problems facing RFID deployments in the 2000s was the belief that any data was valuable data. Companies spent significant money to collect data with no set business goals, leading to disillusionment that likely slowed RFID tagging. In addition, as many scholars have pointed out, the benefits of identifying patterns in data can be overstated when interpretation is ignored. After all, large data sets lend themselves to "apophenia," which refers to "seeing patterns where none actually exist, simply because enormous quantities of data can offer connections that radiate in all directions."[22] For example, Internet Explorer usage and teen pregnancy rates are strongly correlated, but they obviously have little to do with each other. Finally, data can be divided into "captured data" and "derived data": "Captured data are often the input into a model, with derived data the output."[23] In other words, data must often be filtered and turned into derived data to provide a usable output. As discussed later in this chapter, the difference between captured and derived data is an important piece for understanding RFID data that needs to be filtered and "cooked" through software. Even when systems automatically act on data inputs, someone with subject matter expertise created the parameters of when and when not to act.

By no means do criticisms of big data mean that data is not important. Big data is not good or bad. Rather, data can be collected well and analyzed productively, or the entire process can be a mess. The criticisms do, however, serve as a reminder of the limits of data analysis. Data still needs to be interpreted; it still needs to be collected properly; it still needs to be paid for and shaped within larger legislative frameworks. Namely, data is influenced by much more than just the phenomenon it measures, a point I expand on throughout the examples of RFID's role in data collection.

RFID Data and Where to Find It

One of my goals in this chapter is to do more than just look at data, whether big or small; rather I want to peel back some infrastructural layers to show just where a significant amount of data comes from. I also want to expand on the overall focus of the book by looking at how data in these examples powers the infrastructures of identification that shape everything from the food we eat to the way our jeans end up in nearby stores. Consequently, the discussions of identification data in this chapter build on concepts explored earlier and capture the role RFID plays in sorting and differentiating objects in the physical world. First I examine specific uses of RFID to monitor everything from human bodies to cows to pharmaceuticals; I then move to the individual level by looking at how RFID data has been particularly criticized by certain evangelical Christian communities as a form of binary identification data. As the examples show, RFID provides a voice in data projects, but that voice is tempered or amplified by a host of factors external to the microchips, antennae, and readers strewn throughout the physical world.

Tracking and Controlling Bodies

A week before their big trip to Disney World, a family of four receives their MagicBand Disney bracelets in the mail. When they arrive at the Orlando, Florida, airport, they move their bracelet past a reader and board a shuttle to one of the park's hotels; their tagged baggage follows them to the hotel. Once they arrive at the hotel, they use their bracelets to check in to their rooms. They then head out to the park, leaving their credit cards and cash behind. Once they arrive at the park, they wave their bracelets in front of the turnstile and are admitted. The photo taken during their ride on Space Mountain is automatically linked back to their account. The parents then take their children to get their picture taken with Goofy; when they get there, Goofy already knows the children's first names. The family then goes to eat dinner at the Be Our Guest Restaurant and is greeted by name, even though no one ever asked. They sit down and the food they preordered online is brought to them even though they do not know how the waiter knows who or where they are. They pay for their meal by passing their bracelet in front of a reader on their way out. When they get back to the hotel, the door senses their presence and unlocks as they approach.

The scenario described above is not a futurist's dream of wearable technologies. Instead, it is a description of how Disney's MyMagic+ suite of technologies is designed to work.[24] MyMagic+ centers on the RFID-enabled wristband described above. Before people who use the system leave for their trip to Disney World, they receive the MagicBand in the mail. They link the MagicBand to their online account, and they can go online to set preferences, enter payment information, place food orders, and so on. Once they arrive at the park, the MagicBand serves multiple purposes. It works as their payment system in lieu of credit cards or cash, it authenticates their access to the park and enables entry to hotel rooms and rides, and it locates and identifies people as they enter certain areas.

Disney invested $1 billion into the system and began pushing MagicBands in 2013. The band includes active RFID, and the system includes a large infrastructure of readers designed to make Disney visitors feel as if the park actively responds to their presence. The MagicBand is a particularly ambitious example of the Internet of Things, but it also implicates issues of data. Through the MagicBand, Disney is able to create a comprehensive data set in real time about people's movements and purchases in the park. To make the data more useful, Disney even labels each MagicBand with the name of each member of the family. Assuming people wear the correct band, Disney receives more than general mobility and purchasing data: the analysis can also match movement patterns to the demographics of each user, meaning the data produced by a bracelet assigned to a forty-year-old man can be differentiated from the data produced by a six-year-old girl. Potentially, Disney can use the data to identify correlations between behaviors tracked by the MagicBand. For example, maybe people who take certain routes through the park spend more money. Maybe waiting time correlates with expensive visits to take pictures with Disney princesses. Through the extensive data profiles of the MagicBand, Disney gains visibility into consumer behaviors. Ultimately, the MagicBand works as a way for infrastructure to uniquely identify and differentiate the crowd of bodies moving through the park.

Professional sports are another area in which RFID is used to produce corporeal mobility data. Over the last two decades, many sports have begun to use big data analysis, and RFID technology has been used to produce fine-grained mobility data by tracking players' movement. The National Football League (NFL) is one example. Beginning in 2014, NFL teams began outfitting stadiums with RFID readers.[25] Players have RFID chips embedded in their shoulder pads, and the readers track the speed of players by pinging the chips. A small fraction of this data was shared with fans in graphics that showed how fast players moved across the field. Most of the data was used by teams to correlate performance with other factors and enable increasingly advanced forms of data analysis.

While the NFL may be currently engaged in the largest-scale RFID project in the sports world, RFID has been used to produce data about other sports as well. Soccer teams all over the world have used the technology to track individual players' movements to assess speed, limit players coming back from injury, and provide analytics about how players move as a team.[26] Hockey teams in Finland and Canada have used RFID to produce data about players and spot correlations that might identify increased injury risk.[27] Finally, as an example of the velocity of RFID data, marathons have used RFID attached to runners' shoelaces to track split times and share data with the crowd in real time.

A final example of bodies and data—the Sabarimala pilgrimage in India—shifts the focus from the "big" to the individual. The Sabarimala pilgrimage is one of the largest pilgrimages in the world, involving around fifty million Hindu travelers a year. During the pilgrimage, children are often separated from their parents. Starting in 2017, the Kerala State Police now require every child under age fourteen to register and wear a lanyard that features an HF RFID tag that features the "child's names and the guardian's name, contact number and other relevant details."[28] The system was developed by Vodafone India, and the tags work as an infrastructure of identification used to sort individual children from the millions of children moving through the physical space. Here the focus is not on the aggregation of data for analytical purposes. Rather, the focus is on "small" data attached to each child that can play a role in reuniting lost children with their parents.

These cases are specialized examples of the types of mobility data frequently collected at the city level. As discussed in the previous chapter, RFID plays a major role in toll systems and public transportation by connecting objects with infrastructure. In concert with other sources, such as road sensors and GPS-enabled services like Waze, these RFID systems produce data cities can use to monitor flows of movement.[29] In Yinchuan, China, for example, RFID tags were a part of the larger development of the city's "smart" capabilities. The city distributed RFID tags to hundreds of thousands of private vehicles to "monitor the population's movements and manage the traffic accordingly."[30] Cities can measure traffic patterns and determine how long it takes to move from one toll to another through the unique identification abilities of RFID. And, in a few more controversial cases, cities have set up RFID tracking systems that extend past toll stations. As discussed in more detail in chapter 6, New York City built a system of RFID readers used to monitor transportation patterns that pinged people all over the city without their knowledge.[31]

Those flows of movement can provide insights to more than city planners; they can also be visualized to provide a unique perspective to the general public. One set of cases can be found at the Flowing City project, which provides "urban data visualizations of the city, making the city smarter with data."[32] The RFID section of the project displays visualizations of three sources of RFID mobility data, including two visualizations of Oyster card usage in London. The second Oyster Flowprint is particularly interesting and uses RFID Oyster card data to show charts of London Tube ridership and data on the routes people take on the Tube. Through the visualizations, what could be incomprehensibly large flows of RFID data become a way for people to rethink their city and understand mobility patterns. Just like with the data visualizations on NFL or soccer broadcasts, Flowing City shows how an extra level of visual mediation can turn RFID data into actionable, insightful information.

A Disney MagicBand, an RFID tag embedded in a player's uniform, or a subway card all track different types of mobility. The MagicBand and the sports examples rely on a large network of readers, building a surveillant's dream environment, in which longer-range RFID tags attached to individuals are read constantly. The subway card or toll tag logs data only at points of entry and exit. However, despite their differences, all these systems use RFID technology to wirelessly produce data about corporeal mobility patterns, and they are representative of larger trends in the datafication of mobility that have shaped our world. Writing about those trends, French philosopher Gilles Deleuze discussed how Western society has moved away from the "disciplinary societies" analyzed by philosophers such as Michel Foucault.[33] In disciplinary societies, power is manifested through enclosure and boundaries, through limitations on and active surveillance of movement. Deleuze tracked the shift to what he called "societies of control," in which people are allowed the freedom to move but are controlled through data. As Deleuze said, "What counts is not the barrier but the computer that tracks each person's position."[34]

MagicBand, toll tags, and subway cards all play a role in the practices of control. All those technologies allow movement rather than discipline and restrict it. They open up enclosures—whether that enclosure is a tollgate or a part of an amusement park—that they themselves would discipline and restrict behavior. But the price of that movement is the production of identification data. Of course, for Deleuze's societies of control to work on a large scale, the bodies that move must be differentiated. They must be tracked individually before ever being compiled into the aggregate. The examples of RFID covered above provide the identification that powers practices of control: they enable movement and make enclosures porous and malleable, but the price people pay is the tracking of their mobility through RFID.

The simultaneous enabling and tracking of mobility through RFID data shows both the individual and aggregate effects of the technology on corporeal movement. On the individual level, an RFID wristband or subway card transmits mobility data that is logged in the system; but the movement also does not happen without that data because a turnstile will not open or a restaurant order might not be read. So the data is used as a form of control on an individual level by authenticating certain behaviors from certain bodies. On the aggregate level, the data can then be transformed into something else entirely, obfuscating each individual movement as part of the larger analysis searching for correlations and patterns. But, regardless of whether the data is discussed in the aggregate or the individual, the data focuses on identification. The tags are used as a process of sorting one body from another in large networks of communication and identification.

The differentiation of bodies in the environment is a process of control, a process used to create increasingly detailed profiles about previously ephemeral practices. The profiles might be extremely specific, looking for certain patterns that lead to injuries or consumption behaviors, or the data might be used for higher-level purposes to track travel times or popular routes. Regardless, the RFID in these cases turns individual bodies into identifiable data points that communicate with built infrastructure and contribute to new types of visibility about mobility.

Data in Networks of Supply

One of the main growth areas in the RFID industry involves tagging items to be tracked through supply chains and monitored on store floors. As discussed in chapter 2, the push toward item-level tagging contributed to both the peak of RFID's hype cycle and the bottom of the technology's "trough of disillusionment" in the 2000s. RFID at either the pallet or item level is a prime example of big data. Tags produce a large volume of data that can affect visibility and inform decision-making processes. But the data produced by RFID also provides concrete examples of why criticisms of big data are important. RFID is a good use-case of what happens when data is valued purely for data's sake, without recognition of the layers of human intervention necessary in most big data projects.

RFID works as a newer version of what James Beniger called "control technologies" by producing data used to monitor increasingly complex flows of objects through the global economy. Successful tagging projects increase data visibility in supply chains and enable more accurate inventory counts on store floors. In the cases of complex objects (such as cars), individual components often feature their own tags. Objects become more trackable through the large amount of data the tags produce. What might have been long gaps of visibility in a process before, or merely not a particularly accurate data profile, can now be more robust with the embrace of the big data produced by RFID. For example, the accuracy of inventory counts with barcodes is often around 66–70 percent, but that number can increase to greater than 95 percent with RFID technology.[35] But an interesting aspect of some control technologies, including RFID, is that the large amount of data can lead to a control crisis of its own.

To explain why, it is first important to note that item-level tagging can produce far more data than is useful, what Rob Kitchin calls auxiliary data.[36] When passive tags come in contact with readers, they transmit their information constantly until they leave the reader's field. Much of the time, all but one of those reads is useless data.[37] For example, if a distribution center uses RFID readers to log the pallets that move through the system, only one read is necessary to detect an objects' presence. The other tag reads need to be filtered out. The same goes for item-level tags on the store floor. If the system is designed for inventory, multiple reads of the same tag serve little purpose. In addition, RFID tags can contain larger amounts of data, only some of which might be necessary to a specific system. The data in its entirety can quickly become overwhelming and unusable, creating a mini control crisis that must be addressed through further forms of mediation.[38]

Relatedly, some of the problems with RFID deployments also concerned misconceptions about the value of data. Namely, too often companies embraced (or were forced to embrace) the data produced by RFID without thinking critically about that data. As articles in industry journals pointed out, adding RFID to items without a viable business case often does not accomplish much.[39] Without a plan on what the data was for, just collecting massive amounts of data did little to improve business processes and provided little return on investment. Data is not worthwhile just for data's sake. Another problem came in the "big" nature of RFID data. As explained above, RFID produces far more data than almost any business could use. To act on RFID data required complex filtering techniques through software specifically designed for RFID. The software, which is covered in a regular feature in the *RFID Journal*, is itself a further control technology that responds to the data saturation created by RFID tags.

The software as filter is not a minor point about RFID's role in the supply chain. One of the oft-repeated claims about big data imagines that "with enough data, the numbers speak for themselves,"[40] but the struggles of RFID tagging show why those claims can misunderstand the nature of data. RFID data can be overwhelming and messy, improperly structured, and mostly irrelevant. Only after the data is interpreted—first in how the data is structured on the tag and then at the level of software that decides what to present to the end user—does it have value. For example, software filters out multiple reads, software decides whether or not to analyze only part of the data, software visualizes data in a useful way, and software sends alerts on store floors when a product needs to be stocked. Software turns the captured data into derived data that becomes a useful output. The importance of software may seem obvious, but that importance is often lost in discussions about data speaking for itself. RFID technologies are an example of the layers of human interpretation, some mediated by code and some not, necessary for data to work as a secure base of the knowledge pyramid.

Building on the point about interpretation and filtering, a key to the usefulness of RFID in the supply chain was the development of the Electronic Product Code (EPC) (covered in chapter 2). The EPC works like many other standards and provides a structure companies can follow to align data formats.[41] The EPC helps turn tagged data into structured data that can be accessed and analyzed across multiple systems, in contrast to proprietary data formats that limit the impact of RFID tagging. Imagine one company makes pairs of jeans that sell at Walmart and Marks & Spencer. Without the standardization of the EPC, that company would either have to use a different RFID data format for products shipped to each of the two companies, or the two companies would have to maintain an unwieldy database of different tagging standards of their many suppliers. The data is shaped through the intervention of the standards organization before it makes it onto the RFID tag. A process that seems rather straightforward on its face—the object is either there or not there—is quite complex and shaped by decisions made by human actors (the members of EPCglobal), who are far removed temporally and spatially from a distribution center or store floor.

The amount of supply-chain data produced using RFID has increased in recent years, with item-level tagging finally achieving some of the potential predicted in the early 2000s.[42] The increased success is due in part to the lower price of tags, improved technology (e.g., tags that are more reliable), and better software to filter data. But the increased adoption is also due to the better understandings of data discussed throughout this chapter. At the peak of its hype phase, many companies rushed to adopt RFID without establishing how they would use the data. Most of those projects failed, with Walmart's mandate being the most public, in part because Walmart did not make clear how suppliers would benefit from the increased data provided by RFID.[43] Now some companies have established ways to use the data productively, which has led to growth in tagging projects. Here I want to look at a few notable examples of how the ability to uniquely identify objects through RFID has shifted data practices.

The first example involves Delta Airlines' (an early proponent of RFID in manufacturing) tagging of all checked baggage. In

2016, Delta announced it would spend $50 million converting its baggage process to RFID. Other airlines had experimented with RFID, and airports in Las Vegas and Hong Kong had shifted to RFID systems, but Delta's project was the most comprehensive. The project has been successful and, according to Delta, has improved baggage tracking to a 99.9 percent success rate.[44] Possibly the most interesting shift, however, has to do more with the velocity than the volume of data. Delta's system tags each checked bag with a UHF RFID tag. The tags are then automatically read by readers placed at various points in the baggage process, and customers with the Delta app receive real-time notifications about the location of their bags. Through RFID, the visibility of the checked bag's journey alters, and the journey becomes accessible to customers with real-time updates. The velocity of the data is transformed through the comprehensive system built by Delta.

Another case can be found in the men's clothing company UNTUCKit. The company is piloting a project that involves tagging fifty try-on shirts in each store and recording each time the shirt is taken into the dressing room.[45] The tags also produce data when the shirts are purchased, so the company is able to analyze analytics that compare the percentage of shirts people try on to those they eventually purchase. The velocity of data also changes, and when a shirt is purchased, the employees receive an alert that the specific shirt in that size needs to be restocked. Because the tags differentiate clothing to the point that each individual shirt is identifiable, the analytics are more detailed and can be used to record consumer practices in new ways.

One final example covers an area slightly outside the supply chain, but it is beer, so I wanted to include it. A rather surprising area of growth for RFID involves monitoring inventory in beverage services, particularly beer. One of the most prominent companies to do so is iPourIt. The company specializes in RFID-enabled tap systems for brew pubs and provides extensive data about purchasing patterns. At a bar with iPourIt, when customers enter the establishment, they are given an RFID-enabled wristband associated with their account. The customer then walks up to the tap for the craft beer of their choice, scans their bracelet, and pours. The amount of the pour is recorded and charged to the account. The system can also set limits that will stop pouring for customers after they consume a set amount of beer. Data about each pour are recorded in real time, so, to quote the iPourIt website,[46] the system produces analytics reports about

- Daily sales based on product style and customers' gender
- Ounces sold by the hour
- Daily inventory reports
- Discount reports
- Beverage sales breakouts
- Comprehensive sales reports

The iPourIt example shows how the data produced by RFID can contribute to new forms of visibility. With traditional bar systems, beers are not recorded to individual consumers besides through credit cards or tabs operated by bartenders. The inventory is not nearly as exact because individual pours are not tracked to the ounce or milliliter. By automating the process and linking identified bodies to the system through RFID, practices of beverage inventory are fundamentally shifted through data.

The data produced through RFID has not necessarily changed that much in the past few years. Tags were always capable of transmitting unique IDs and associating practices with specific accounts. What has changed is the understanding of how to use the data for informed decision making, an important point that pushes back against some of the more utopian rhetorics of data as a social or business good in itself.

Livestock, Data, and Crisis Control

Except for what I purchase directly from community markets, I have almost no idea where most of my food comes from. The food chain is a black box to me, and I doubt I am alone in my ignorance. Most people do not know where the cows that become their beef are raised, and they likely have little knowledge about the processes of food traceability or the data that agencies rely on for managing risk. As examined in this section, RFID plays a role in producing identification data about animals throughout networks of food supply.

One of the first uses of RFID technology was to tag livestock in the 1970s, and tagging has increased significantly since then. In 2017, 480 million animals were tagged with RFID, in part because of food regulations enacted over the last decade and a half.[47] For example, the United States Bioterrorism Act of 2002 in the United States dictated that when food is

> inspected at ports of entry into the United States, the submission to the Secretary of a notice providing the identity of each of the following: The article; the manufacturer and shipper of the article; if known within the specified period of time that notice is required to be provided, the grower of the article; the country from which the article originates; the country from which the article is shipped; and the anticipated port of entry for the article.[48]

The Bioterrorism Act is one of many regulatory acts that require fairly extensive data about food shipments. That specific act did not mandate RFID technology, but it contributed to the slow growth in using RFID to track shipping containers to authenticate incoming food shipments.[49] RFID, unlike other auto-identification technologies, has the storage capacity to hold a variety of data to identify points of origin and temporal information about food products. In addition, RFID has a long history of being used in shipping, and the New York Port Authority was one of the first agencies to experiment with the technology back in the 1970s.[50]

Regulatory agencies in the United States have most often focused on recommending but not requiring specific auto-identification technologies, but other nations have required RFID to track data about parts of the food supply chain. Canada is a prominent example, requiring all cattle to be tagged with RFID for identification purposes.[51] The following statement from the nonprofit Canadian Cattle Identification Agency (CCIA) shows the scope of the project:

> CCIA has allocated more than 65 million unique numbers, making the CLTS one of the most comprehensive national animal identification systems in the world. Today with over 350,000,000 animal identification, premises identification, animal movement and age verification events in the CLTS, the database actively serves its clients as the largest source of traceability data in Canada.[52]

Canada is not alone in its reliance on RFID to manage livestock. Australia's National Livestock and Identification System also requires all cattle be tagged with RFID that contains a property identification code that links cows to the location they were initially purchased.[53] In addition, some US states require cattle to be tagged with RFID.[54] The voluntary National Animal Identification System (NAIS) in the United States, though it does not require a specific technology, identifies RFID as a recommended identification technology.[55] In sum, many countries now either require RFID or at least recommend RFID be used to track cattle in the livestock food chain, and RFID is not exclusive to cattle. The European Union requires all sheep and goats to be tagged with RFID,[56] and pig ranchers in the United States have the option of replacing traditional ear tags with RFID.[57]

So now the question is why. The answer is the complexity of the livestock supply chain. Livestock is shipped all over the world. In the United States, around 40 percent of cattle sold each year is shipped between states, and the United States exported over $6 billion of beef in 2016. EU nations have a fairly open market for animals and also export animals outside the European Union. In addition, India, Australia, and Brazil are the world's three largest beef exporters.[58] Without extensive data tracking shipments of livestock, it can become difficult to trace points of origin for shipments. In other words, without identification technologies, the food chain can become a black box even to regulatory agencies.

That potential black box needs to be opened up in moments of crisis. Maybe the most famous example of crisis in livestock is bovine spongiform encephalopathy (BSE, or as it is colloquially known, mad cow disease). The United Kingdom faced major BSE outbreaks throughout the 1980s and 1990s and had to destroy 4.4 million cows at a cost of around £5 billion;[59] consequently, the United Kingdom developed fairly advanced animal tracking regulations that suggest RFID as a secondary technology to identify points of origin for infected animals. The United States and Canada, on the other hand, had little data visibility on the flows of livestock in the early 2000s. Consequently, when Canada faced an outbreak of BSE in 2003 and another in 2005 (the first was 1993), the government mandated the RFID requirements discussed earlier.[60] The first case of BSE in the United States occurred in 2003, and the outbreak exposed the lack of data available to the US beef industry:

> The larger issue of debate, however, dealt with the United States' lack of a national animal identification system, which would aid in the location of other animals that may have been exposed to BSE. The media further reported and warned that it could take weeks, or even months, to discover where the Holstein cow was born, and without a national animal identification system, investigators may never know.[61]

The inability to track the origin of the affected cow in the 2003 case—it was eventually traced back to a cow imported from Canada—was blamed on the lack of a national animal identification system (NAIS). Following that case, the US Department of Agriculture claimed that developing a NAIS was a national priority. However, more than a decade later, the system is still only voluntary, despite initial goals to slowly transition to a mandatory system. Though the NAIS has refused to mandate use of a specific technology, a 2005 Department of Agriculture working group recommended RFID be adopted by ranchers.[62] Of course, as mentioned earlier, other governing bodies have gone much further in requiring rather than recommending RFID technology.

Cows might seem like a strange area of focus for a discussion of RFID and data. However, the growth of RFID in livestock management shows how infrastructural technologies few people are aware of affect their daily lives. Many people who eat beef have little idea where the beef comes from or how the supply chain is managed. But the data produced through RFID technologies enables cows to be uniquely identified and communicate their point of origin. The data are particularly useful whenever products need to be recalled because the data trail can lead back to specific sites of origin. Millions of cows move from state to state and from country to country. The cows must be differentiated from one another to respond to crises of control, and the "mass" of animal movement must become individualized and trackable.

That tracking and individualization is built on infrastructures of identification such as RFID, in essence turning each cow into an early version of Bruce Sterling's spimes. The data on the tags can contain the cow's identity, the point of origin, and possibly other locations in the cow's journey. Each tag then has a unique story to tell, and those data stories rise to crucial importance at moments of recall and crisis. But for the stories to exist in the first place, the livestock first must be sorted from all of its kind, which is the crucial role RFID plays as an infrastructure of identification.

The RFID tags also show how data and visibility are more complicated than just technology and ID numbers. Rather, in the case of livestock tracking and identification, crisis and regulation played a major role in spurring RFID adoption. Canada adopted RFID after a BSE outbreak. Australia adopted RFID to address regulations and "to facilitate and ensure export access to European markets."[63] France mandated RFID tagging of goats and sheep because of outbreaks of disease.[64] The data itself may have been beneficial without regulation, but it would have been difficult to get individual ranchers to invest in RFID at a significant scale. External factors—more than the spread of a superior technology or a vague move toward a "big data revolution"—led to the push for increased data collection and the creation of national systems for animal identification. The data represents more than a single technology or individual data points; instead, the growth of RFID adoption in livestock responds to an existing crisis of control and shows why data is never solely about the technologies used to produce it.

Cold Chains and Sensors

The shifts in data practices are not just about the "big." As explained earlier, "big" data existed long before the new millennia. Nations had censuses, organizations managed large amounts of barcode data, companies and governmental organizations monitored food safety data. One of the major shifts with the growth of new data practices, however, was the second V in the three Vs: velocity. Data can now be collected and acted upon much faster, in some cases through the wireless communication capabilities of RFID. The velocity of transmission can enable people to make decisions in real time rather than make decisions based on retrospective data. One example can be found in the growth of RFID in what is called the "cold chain."

The cold chain is a temperature-controlled supply chain. A bit confusingly, a cold chain does not have to involve products staying cold. Instead, it is the process of monitoring whether a product stays within a certain temperature band (whether warm or cold) throughout the supply chain. For example, milk and fish have a relatively small temperature band in which they stay safe and fresh. Many pharmaceuticals also must be kept within relatively small temperature bands to retain their efficacy. RFID is an increasingly popular choice to monitor temperature bands, and RFID-based cold chain technologies are expected to be adopted more widely over the coming years, with one forecasting report predicting a combined annual growth rate of 29.1 percent through 2022.[65]

Cold chains were monitored long before RFID, commonly through what are called "temperature strips."[66] However, RFID impacts both the volume and velocity of temperature data collection. With temperature strips, people often removed the strip after a shipment was complete, accessed the data, and determined whether a product stayed within the proper temperature range.[67] The data was collected manually and used for retroactive decision making because the strip is monitored only at points of arrival. RFID—often active RFID equipped with sensors if the system is designed to log the data—provides a larger amount of data, transmitting temperature readings multiple times per minute to the system. And, possibly more importantly, the velocity of the cold chain alters through the object communication of RFID: because the temperature data is consistently transmitted, systems can set up alerts when the sensors detect temperatures outside set parameters. The data can then inform real-time shifts in shipping behaviors.[68]

The cold chain is an example of how RFID's pairing with sensor technologies enables increased environmental monitoring as part of the Internet of Things. The setup becomes a linking of "things" to produce data about environmental factors and inform decision making based on data outputs. While an RFID technology monitoring the temperatures of a shipment of chicken or pharmaceuticals is smaller scale than the wireless sensor networks discussed in chapter 3, the example represents the types of environmental data production Weber and Wong argue will define the Internet of Things.[69] After all, much of the excitement about the Internet of Things focuses on data. RFID's role in the cold chain is just one example of how the velocity and volume of environmental data can shift through emerging practices of object communication.

Identification and Counterfeit Pharmaceuticals

RFID has also played a role in monitoring objects, particularly pharmaceuticals, through the production of a different type of data: authentication data. Counterfeit pharmaceuticals are a significant problem. The European Union reported that pharmaceuticals are the third most counterfeited item after CDs and DVDs. In 2015 alone, Interpol seized tens of millions of counterfeit medications, and the market for counterfeit medications has increased over the last decade: by 2016 the market was estimated to fall somewhere between $75 and $200 billion.[70] Although counterfeit drugs are a problem in industrialized nations, less-well-off nations face more significant problems with counterfeiting. The International Trade Administration estimates that as many as half of the drugs sold in some low-income nations are counterfeit.[71]

Counterfeit drugs have serious real-world effects. An investigation in India found that eight thousand people died over a five-year period at a rural hospital because a commonly prescribed antibiotic had no active ingredient. More than one hundred patients in Pakistan died because of a counterfeit tuberculosis drug, and 2,500 patients died in Niger after injecting a fake meningitis vaccine.[72] One of the most notable examples in the United States occurred with the blood thinner heparin. In 2008, the active ingredient in a batch of heparin was replaced at a manufacturing plant. The counterfeit product was then prescribed and caused as many as eighty-one deaths.[73] The United Kingdom had problems in the mid-2000s when a large batch of the cholesterol medicine Lipitor was counterfeited and had to be recalled.[74] The problems in lower-income nations are more severe: the World Health Organization (WHO) estimates that one hundred thousand deaths each year in Africa can be attributed to counterfeit drugs,[75] and one WHO survey estimated as much as 64 percent of Nigeria's antimalarial drugs are ineffective.[76]

The problems with counterfeiting can be explained both by the high value of pharmaceuticals and the complexity of the supply chain, which involves manufacturing centers (sometimes multiple for the same drug), distribution centers, pharmacies, relief organizations, and so on. RFID, in no small part because of its ability to make objects uniquely identifiable through data, has been explored as a possible solution for more than a decade. In 2004, the US Federal Drug Administration (FDA) listed RFID as an emerging technology in the fight against counterfeiting. In 2004, Purdue Pharma began a pilot project to tag the heavily counterfeited opiate Oxycontin and expanded the pilot project in 2007. In late 2005, Pfizer began a large-scale pilot project to tag all bottles of the erectile dysfunction drug Viagra sold in the United States.[77]

Since those early deployments, some other pharmaceutical companies have also deployed RFID to battle counterfeiting, though adoption has been slower than initially hoped for two primary reasons: cost and regulation. Estimates suggest implementing RFID across a pharmaceutical supply chain costs roughly $10 million to $25 million USD, though studies suggest the return on investment of replacing existing barcodes and paper pedigree systems could recoup that cost in a few years.[78] Another problem has been the complexity of different regulatory frameworks. Regulations between nations differ greatly, but even within just the United States, pharmaceutical tracking is determined by a patchwork of state regulations.[79] These different regulations can make it difficult to implement wide-scale RFID solutions that can address different rules.

In the United States, the regulatory framework began to partially clear up in 2013 when President Obama signed the Drug Quality and Security Act (DQSA). A subset of the DQSA, the Drug Supply Chain Security Act (DSCSA), established new federal requirements for the track and trace and identification data for pharmaceuticals. Part of the DSCSA focuses on establishing a national electronic pedigree database by 2023, replacing the typically manual pedigrees in place now. An electronic pedigree is an electronic document that displays the history of a batch of a drug. Electronic pedigrees are more difficult to tamper with, enable more advanced track and trace capabilities, and allow for increased interoperability across state lines. Unsurprisingly, RFID tags are an enabling technology of electronic pedigrees. Using the EPCglobal Pedigree Standard, a unique EPC can be assigned to each pill bottle and a connected RFID reader can access the EPC and then test the code against the data contained in either a shared or proprietary database.[80]

The idea of huge pharmaceutical companies getting increased control over drugs is not necessarily a positive in all cases. But I want to note that the difference between generic drugs and counterfeit drugs is important. Generic drugs have the same active ingredients but are sold for lower prices under a different name. As far as I can tell, the use of RFID will not affect vulnerable communities' ability to get generics; the generics would still be legitimate and tied to a pharmaceutical database. Counterfeits, on the other hand, are illegal and often do not contain the same ingredients. They are intentionally mislabeled so the person taking (or distributing) the drug thinks they are taking (or distributing) something else. By definition, just having a drug with slightly different ingredients does not automatically make it a counterfeit. It becomes counterfeit when it is intentionally mislabeled and sold for profit. The RFID authentication systems could help combat those practices.

In few areas is the idea of Bruce Sterling's spime, from chapter 1, more immediately important than with pharmaceuticals. RFID's ability to make each bottle of pills or package of vaccines uniquely identifiable and capable of transmitting the history of the batch has the potential to save lives.[81] However, as important as improved pharmaceutical authentication may be, pharmaceuticals also show the challenges of data collection and production.

Namely, data often cost money to produce and collect. Switching from barcodes to RFID is not cheap and takes time. Even if drug companies all agreed that RFID could better combat counterfeiting, the switch would still be slow. The US FDA, for example, provided a ten-year window just to switch to e-pedigrees.

The example of pharmaceutical authentication also provides a reminder that, despite RFID being well established in some areas (e.g., transportation), the technology still has significant room to grow. The requirement in the United States to mandate more detailed, interoperable electronic records by 2023 has led to some excitement in the RFID pharmaceutical space.[82] But activity in the pharmaceutical sector is a reminder that data costs money, sometimes a significant amount of money. Counterfeiting is a major problem; improved authentication data would help combat counterfeiting. But almost a decade and a half after the FDA first identified RFID as a key technology for battling counterfeiting, the majority of pharmaceuticals still rely on older forms of data transmission for authentication. Only time will tell whether RFID—in many ways a superior technology for producing identification data about pharmaceuticals—wins out or whether already established forms of authentication maintain their position.

Blockchains and RFID

This book has already covered the Internet of Things and big data, so it is not all that surprising that RFID has also been related to an even buzzier tech word: the blockchain. Blockchains are an identification infrastructure of sorts, and they work as a distributed ledger that creates records of transactions that cannot be changed. Blockchain technology was invented to power the cryptocurrency Bitcoin, and Bitcoin is still the most famous use of the blockchain. With Bitcoin, the blockchain allowed the currency to verify transactions in a permanent matter and make sure people did not spend the cryptocurrency twice, all without the oversight of a third party such as a bank. The blockchain was a major part of what made Bitcoin so revolutionary. The distributed ledger created a peer-to-peer identification infrastructure that did not require third-party oversight.

Blockchain technology was originally designed for Bitcoin, but in late 2017 and early 2018, suggested uses of blockchain began popping up everywhere. If one were to buy in wholesale to utopian tech rhetoric, then blockchain technology will supposedly "transform the face of digital democracy,"[83] create "a new future for digital identity,"[84] and maybe even change world production of organic bananas.[85] The hype got so big that companies were adding "blockchain" to their name and watching their stock value double.[86] At one point, the chairman of the US Securities and Exchange Commission warned companies against adding blockchain to their name just to drive up stock prices.[87]

RFID has been linked with blockchain technology to create a double-layered, comprehensive infrastructure of identification. Basically, the data recorded through RFID in a supply chain would then be stored in a blockchain. Each tag read at various stages of an object's journey would be shared to the distributed ledger, verified among the peer computers based on certain criteria, and then added permanently to the blockchain. The chain would then create an immutable record of mobility through the comprehensive tracking data produced by RFID. A *Forbes* article describes one possible use-case:

> Pallets with RFID tags would communicate their need to get from point A to point B by a certain date. Carrier "mining" applications would bid for the right to move that load. The RFID tag would award the business to the carrier that bests meets a shipper's price and service needs. Then as the move progresses, the blockchain would continue to track the shipment.[88]

The RFID in these cases would not change, but the blockchain adds another infrastructural identification layer—the distributed ledger—on top of the built infrastructure of identification of RFID. The data being transmitted remains the same, but the storage and authentication practices shift through the distributed ledger. The extra identification makes it easier to share and verify data among different actors in the supply chain because no third part oversight is required. The blockchain would also be available to all interested parties and cannot be tampered with, reducing the need for different actors to use compatible forms of software. Or, as GS1 phrased it in a press release announcing a partnership with Microsoft and IBM to explore blockchain and RFID, the blockchain would "maintain a single, shared version of the truth about supply chain and logistics events—increasing data integrity and trust between parties, and reducing data duplication and reconciliation." Essentially, the dream of RFID + blockchain is a nearly fully visible, accurate supply chain that can be accessed by multiple relevant parties at once.[89]

Various sources have theorized how to use blockchain technology with many of the RFID cases in this chapter, including retail,[90] food,[91] and pharmaceuticals.[92] RFID even has its own cryptocurrency, called Waltonchain,[93] which might not be as impressive as it sounds, considering there are now cryptocurrencies for everything from Kodak to dentists.[94] At this point, it is difficult to separate hype from reality, and some sources have cut through some of the utopian hype to argue blockchain technology may not be nearly as flexible as evangelists believe. But regardless, the pairing of the two could represent the creation of extensive, permanent, peer-to-peer ledgers of identification information. The two would also represent a layering of infrastructures of identification, with the top layer being RFID technologies and the backend involving the algorithms that power the distributed ledgers and authenticate various transactions. The two technologies are consequently related in the way they shift the increasingly granular production and verification of identification data on different scales.

The Mark of the Beast

As a media scholar, I have read countless articles from tech publications and academic journals. I am used to it. What I did not expect my research to lead me to, however, was a rereading of the Book of Revelation. And yet, in the process of studying RFID and identification data, I ended up having to find a Bible to read prophecies of the coming of the Antichrist. This final discussion explains why I am transitioning from bigger, often aggregated data to RFID data on a very personal, very biblical level.

The Book of Revelation is a book in the canon of the New Testament of the Christian Bible. Revelation is unique in the New Testament as the only book in the canon that deals explicitly with the apocalypse. The book is also unique in style and features much of the apocalyptic imagery that has shaped the Western popular imaginary, including the number 666, the seven-headed dragon, the seven seals, and the four horsemen of the apocalypse. But one particular passage has made the book relevant to discussions of RFID:

> And he [the Beast] causeth all, both small and great, rich and poor, free and bond, to receive a mark in their right hand, or in their foreheads:
> And that no man might buy or sell, save he that had the mark, or the name of the beast, or the number of his name.[95]

The passage above refers to what would eventually become known as the "Mark of the Beast," a mark that works as a form of identification data that differentiates those who worship the Antichrist. To some evangelical Christian observers, RFID has a lot in common with that mark.

The warnings suggesting links between RFID and the Mark of the Beast began appearing in the early 2000s, and many, but not all, focused on subcutaneous RFID. By no means was RFID the first emerging technology to get the "Mark" label. As one evangelical advice column about RFID put it, "You recall the uproar when credit cards became standard fare in America. Many were afraid that the Beast was upon us."[96] But with the growth of RFID in the early 2000s, discussions of the Mark of the Beast increased significantly in some evangelical communities.

RFID has been identified as a possible mark for two primary reasons. The first reason is that subcutaneous RFID is most typically injected into the palm of the hand, and the biblical passage specifically mentioned the hand. The second reason is that RFID can contain payment information, which links the technology to the second part of the passage. Some of the biohackers I discussed in chapter 4 have injected RFID chips that contain their payment data, enacting the type of future feared by the many evangelical Christian warnings about RFID.

More articles than I could begin to cover here have been published on evangelical sites warning about RFID and the Mark of the Beast. For example, in chapter 6 I discuss the book *Spychips*, which is the most prominent mid-2000s book written about

RFID and surveillance. The authors published an alternative version of the book targeted at evangelical communities that added passages describing RFID as a possible Mark of the Beast used to sort people. The evangelical version of the book even had a different name: *The Spychips Threat: Why Christians Should Resist RFID and Electronic Surveillance*.[97] Many other articles online warn against RFID adoption and argue that implanted chips (or even RFID-enabled cards or bracelets like the Disney MagicBand) are a tool of the antichrist and a possible sign of the end times.[98]

The Three Square Market case discussed in chapter 4 is a recent example of how these discourses arise. As a reminder, Three Square Market is a company that offered employees voluntary RFID injections. After the story gained national attention in the United States, the company's online presence was flooded with over a hundred one-star Google reviews, many of which were from evangelical Christians. Most of the one-star reviews focused on the sin of using RFID as a form of identification and payment. For example, reviews say things such as the company is "doing the dirty work for Satan himself" and "People read your Bible. This is the first sign of the mark of the beast." Articles with titles like "Would you accept the MARK OF THE BEAST to get a job?" then began appearing on various religious websites.[99] The Three Square Market case is one of many, but it is a reminder that, in certain communities, RFID has maintained its place as a particularly troubling technology that hearkens back to biblical prophecy.

The Mark of the Beast might not seem particularly relevant to a broader discussion about data, but the mark is ultimately an issue of data. These groups believe people who have the mark worship the beast; those who do not have the mark will be persecuted but eventually redeemed. If you believe in a fairly literal interpretation of the Second Coming, there might not be a more important data point than the "for us or against us" of redemption symbolized by the mark in Revelation. According to the prophecy, the mark becomes a binary identification infrastructure that sorts the population into two groups, with the highest stakes possible. The data in that case might not be "big" or varied, but it is certainly consequential.

The fears of these groups also raise larger questions about the convergence of data within the human body. The framing through apocalyptic imagery is rather unique, but the issues raised extend past the narrow confines of whether people believe in the Antichrist or not. Ultimately, the discourses surrounding the mark focus on questions of identification and payment data moving into the body. Each body becomes uniquely identifiable through the mark, and just as importantly, the microchips in these cases can potentially contain unique payment identification as well. The body in evangelical fears becomes a nexus through which capitalism, broadly defined, must pass in terms of exchange of payment information. The body also, as mentioned earlier, becomes sortable through this mark foretold in the Book of Revelation.

In sum, the case of the Mark of the Beast is a departure from the other examples of RFID as data infrastructure covered above. But these evangelical discourses raise questions about RFID and data on a more specific level that provide a different perspective on the kinds of sorting enabled through the various RFID identification infrastructures covered above. While it would be easy to dismiss these fears, and I personally do not believe RFID is a tool of the Antichrist, they do raise interesting questions about how specific communities understand the convergence of data and flesh. Finally, there is something particularly interesting about the idea of an infrastructure of identification that produces binary data about the most dramatic kind of sorting possible: the sorting of the damned from the redeemed.

The Tags behind the Data

This chapter has focused on RFID and data. However, most of this book is about data. RFID, after all, has the potential to turn almost any object into identification data, and the examples in earlier chapters and in chapter 6 involve data exchange of some sort. A pet or human with an implanted RFID chip sends identification data to a reader; a badge to enter a building interacts with a larger system to authenticate data. And so on. The main purpose of RFID, just like the main purpose of barcodes, comes from the identification process that cannot happen without data. Even the idea of a biblical mark on the hand is a type of data, albeit less complex than millions of tags transmitting EPCs in a supply chain.

The examples covered in this chapter show why RFID fits within the larger growth of data collection practices. The cause and effect relationship of technologies and society is complicated, and the discussion of RFID shows why that is the case. On the one hand, RFID is a key enabler of many data projects. In some cases, the same scale of comprehensive data collection about a phenomenon would not be possible without RFID's communicative capabilities and larger storage capacities. For example, no amount of barcode infrastructure in retail could produce as huge a data set as an equivalent RFID system; paper records of pharmaceutical pedigrees would likely not be as extensive as the e-pedigrees enabled by technologies like RFID; brands or barcodes on cattle are not as potentially comprehensive as RFID ear tags. The sheer amount of data produced in these projects would be difficult to imagine with older techniques of production and collection.

On the other hand, technologies like RFID do not *cause* the larger societal push toward a supposed big data revolution. Instead, part of the hype about RFID can be explained by the increasing recognition of the importance of data. Back in 1999, when item-level tagging and the Internet of Things was first envisioned, RFID was seen as a key enabling technology because it could provide the connective data to power logistical systems. The idea of improving the supply chain through data already existed; RFID was seen as a technology to make improved data visibility a reality. I do not want to belittle the roles that technologies play. The exponential increase in the amount of data produced does not happen without a shift in various technologies. But as shown in the cases discussed above, the insertion of RFID as a mediator between objects and databases was often a response to some kind of existing control crisis: the twenty-fifth anniversary of the barcode discussed in chapter 2, the increasing complexity of supply chains, outbreaks of BSE and regulations on livestock tracking, an epidemic of counterfeit pharmaceuticals, and so on.

The examples also show why big data is more complicated than just collecting data and letting it speak for itself. The three Vs of big data—volume, velocity, and variety—can all be improved by RFID. An RFID system can produce far more data than a barcode system; can be read faster than a barcode; and can contain different types of data, especially when paired with sensors. But collecting data and choosing what to collect is a complicated process shaped by various factors. In the mid-2000s, the Walmart mandate mostly failed because suppliers saw few benefits, had few plans on how to use the data, and felt forced into using RFID by a corporate giant. The technology also often cost too much, though tag prices have come down since then. With livestock, RFID became a requirement in some countries only after they faced the crisis of BSE. The United States still only recommends RFID, a decision likely shaped in part by the US government's historical fear of overregulation. With pharmaceuticals, issues of cost and patchwork regulations have slowed RFID deployment, though many sources agree the technology can provide more complete identification data to improve pharmaceutical traceability.[100] Finally, the Mark of the Beast discussion shows how certain data practices are linked to much larger issues; no matter how useful RFID data becomes in the future, certain groups of evangelical Christians will likely never willingly adopt the technology.

Conclusion

Part of what I hope to have shown in this chapter is that data projects are about more than data, about more than technology. That is not to say that technological efficiencies do not matter. At the peak of the RFID hype in the mid-2000s, tags sometimes did not work as well as advertised and had problems with read collisions, water, and so on. But, more than just the efficacy of data-producing technologies or the ability to collect identification data, factors external to the agential capabilities of RFID have shaped the technology's role in data projects: software has improved, which makes filtering data easier; the EPC standard gained widespread adoption and provided an established data format; the price of tags decreased and companies developed more concrete plans on how to use the data; and regulations have either helped or hindered adoption of RFID across industries. In sum, the data produced through RFID is made valuable once factors external to the technology fall into place.

The value of external factors is why I discussed the pairing of blockchain technology and RFID. RFID data can be massive and often needs to make sense to multiple audiences. In a supply chain, for example, multiple organizations might need to be able to access shipment data about an object. Blockchain works as a type of shared (not copied) database that can securely record data while enabling different actors to contribute RFID reads to the ledger. The blockchain is often discussed in terms of transparency and visibility, which coincides closely with the business discourses about using RFID to improve supply-chain visibility. Blockchain technology could become increasingly central to various RFID data practices, though various problems—such as the massive amount of power consumed and relatively low speed of the blockchain—could hold the pairing back.[101]

Acknowledging the external factors that shape RFID's role in data projects does not mean the technology itself is not worth discussing. As Katherine Hayles argued, the billions of RFID tags moving through the world are important in their own right.[102] They are active agents that produce meaning and show how our world is shaped by communication that occurs beneath the level of human perception. After all, unlike barcodes, which must be scanned, the examples covered in this chapter represent the transmission of data that can occur without direct human intervention, at least once the systems are in place. Take the Disney MagicBand as an example. Someone moving through Disney World shares their identifying data whenever they come in contact with a reader. They transmit data about movement possibly without even knowing what their wristband does. Or take a shipment of fish in a cold chain that uses RFID and sensors. The temperatures are transmitted wirelessly throughout the journey, and no human has to manually record those readings.

RFID is just one of many data-producing technologies. In many ways, it is a simpler part of the overall data landscape than other areas one could examine. An object that shares identification data through a tag is less complex than much of the machine-to-machine communication occurring in artificial intelligence projects, less massive than data shared over internet or mobile phone networks, less flashy than advanced social media analytics. The vast majority of objects with RFID—whether a cow, a pair of jeans, or a bottle of pills—have no computing power of their own. They are entered into networks through attached tags, and often the tags themselves do not even have an internal power source. That simplicity, however, is part of what makes RFID technologies so interesting. Tags can attach to almost any object and make it machine-readable. They can transmit data in near perpetuity because they have no internal battery. And they can do all that for as little as ten cents a tag. Returning to the image in chapter 1 of tags I collected in the course of my research, each one of those tags, in all their different shapes and sizes, can turn a body or an object into uniquely identifiable data. These tags, some smaller than a fingernail, are what makes RFID such a crucial contemporary infrastructure of identification.

6 Surveillance and the Mobility of Bodies

Something about the word "data" seems sterile, the implication being that data is something objective and cold stored in spreadsheets. But data always describes something. Some types of data, such as sensor data about the environment or data that tracks objects in a supply chain, may not raise many concerns about personal privacy. Much RFID data, however, is about people. RFID technology logs people's movement on public transportation, transmits their data on ID cards and passports, tracks their movement in school and work environments, and can be sewn into clothing or implanted in bodies.

For those reasons, RFID technology has repeatedly been the target of privacy advocates. Protests and boycotts pushed back against the technology in the mid-2000s,[1] the US Federal Trade Commission and the European Commission formed working groups to deal with RFID and data privacy,[2] and books were published that compared RFID to schemes from Nazi Germany.[3] In sum, RFID struck a chord among privacy advocates, and protests shaped some of the public perception of the technology in the mid-2000s. These protests even led some companies to back away from the RFID label. As discussed earlier, the Smart Card Alliance—a trade organization for contactless payment cards—features an FAQ insisting their technology is not RFID, despite the fact that smart cards use RFID.[4] People I interviewed told me companies developed pilot projects that use RFID tagging but did not make the projects public because of fear of backlash. This chapter traces the development of RFID privacy debates and shows how public concern likely peaked around 2007 and then tapered off, even as RFID has finally begun to meet some of its earlier potential.

Privacy and surveillance are two of the most well covered areas in media studies. Scholars have written extensively about how privacy has shifted with the growth of social media,[5] how locational privacy issues are affected by mobile telephony,[6] how companies abuse data collection practices online,[7] and how various urban technologies such as CCTV are used to surveille the population.[8] When talking about privacy and surveillance, it is first important to note that RFID is merely one of many technologies that can be used to monitor the population. As sociologists Kevin D. Haggerty and Richard V. Ericson argued, any discussion of a technology of surveillance must be contextualized within the larger "surveillant assemblage" that combines data from different sources.[9] After all, there is no Panopticon with one large window through which institutions observe the public. Instead, contemporary surveillance works more like the vision of a fly, broken into many related windows on the world.

But if we live in a world with many windows, RFID has the potential to be a fairly clear pane of glass. RFID, after all, is a technology of communication and identification, a technology that can be used to turn many physical processes into analyzable data. Many of these processes represent mundane actions such as walking into a classroom, but as surveillance expert David Lyon argues, "surveillance occurs in the most high-tech ways and at the pinnacles of power but depends on the humdrum, mundane communications and exchanges that we all make."[10] The often "humdrum, mundane communications" discussed below are not necessarily good or bad, and this chapter examines concerns expressed about RFID while avoiding the dystopian claims that shape much of the privacy discourse that surrounded the technology. But first, it is important to remember just why some people fear RFID so much. To get an idea, one can turn to a 2014 European Commission report that argues the privacy concerns "have been fueled by the specific characteristics of RFID:

- RFID tags and readers are largely invisible, allowing information exchange without the individuals concerned being aware of it;
- Tags are always-on and can usually not be switched off;
- The technology is expected to become increasingly pervasive."[11]

These three issues will come up throughout this chapter. In short, unlike with their mobile phones, people may not know they have come in contact with RFID. Unlike the internet, people cannot close a browser window and stop transmitting data. RFID tags tend to be hidden, infrastructural technologies that move as people move, passively producing identifiable traces of mobility.

The three issues identified by the European Commission also show how RFID technology is part of the growth of what David Lyon has labeled "liquid surveillance."[12] Building on philosopher Zygmunt Bauman's work on "liquid modernity,"[13] Lyon has shown how surveillance is increasingly mediated and automated by software. Unlike older forms of surveillance that involved human intervention to watch people, to record data by hand, or even to choose where to place a body to surveille other bodies, contemporary surveillance increasingly happens in more "liquid" forms that do not require much human intervention. The examples I discuss below do not involve people watching or even people reading through spreadsheets to note logged data. Instead, whether the example involves using a card to get through a subway terminal, a badge to get into a building, a bracelet to automate attendance, or a passport to cross a border, the data traces are reliant on software and hardware. RFID, after all, plays a central role in animating the physical environment in new ways, and it does so while working as a comprehensive infrastructure of identification. Consequently, this chapter will show how practices of surveillance have become increasingly automated through the data collection of RFID technology.

Understanding RFID as a piece of the sociotechnical surveillant assemblage requires more than just an analysis of the technology. Consequently, I first trace the trajectory of RFID's position in the public eye, especially focusing on protests that roiled the industry in the mid-2000s. I then move on to different areas of concern with RFID and surveillance, including general mobility, retail surveillance, workplace and school tracking, biometric identification, and bodily implantation. I conclude by situating RFID within larger discourses about individual privacy and suggest a possible way forward that embraces the "infrastructural imagination" that encourages people to examine their environment in new ways.

Boycotts, Privacy, and RFID

Privacy has many definitions, but the one I use throughout this chapter is media scholar danah boyd's definition of privacy as the "control over social situations."[14] For people to maintain a sense of privacy, they must be able to control how their information is accessed through both technical and social means. Another useful step involves splitting the umbrella term of privacy into two separate spheres: social and institutional privacy.[15] Social privacy focuses on information sharing among people on a relatively equal social level. Institutional privacy concerns information sharing among individuals and larger institutions such as governments and businesses. As the history of debates about RFID and privacy suggests, the major concerns are with the institutional side of privacy; advocates in the mid-2000s argued that people had little control over the information shared with institutions through RFID.

The privacy concerns about RFID began not long after Kevin Ashton first gave his presentation on the Internet of Things in 1999. In the initial period after the presentation and the formation of the Auto-ID Center, the technology remained obscure and received little public attention. The initial period of obscurity—what the Information Technology and Innovation Foundation calls the "trusted beginnings" stage[16]—only lasted until around 2002. In 2002, one of the major players in this story entered the scene: Katherine Albrecht. Albrecht is the founder of Consumers against Supermarket Privacy Invasion and Numbering (CASPIAN), a privacy advocacy group that had an impact on the RFID industry in the mid-2000s. CASPIAN was created in 1999 as an organization dedicated to fighting against supermarket loyalty cards. However, the fight against supermarket loyalty cards did not gain significant traction, and the organization began to focus on RFID in the early 2000s.

One of the first major strikes CASPIAN organized against RFID tagging came in 2003. Earlier that year, Philips Semiconductors partnered with the clothing retailer Benetton to use RFID to track clothing that was part of Benetton's Sisley line. CASPIAN, which had already created a separate website that focused on RFID,[17] organized protests and a boycott against Benetton. While there was some confusion about whether Benetton actually planned to use the tags or was just running a trial program, the protests seemed to work, and Benetton backed away from its RFID project.[18] A similar protest organized by CASPIAN targeted Walmart and focused on a "smart shelf" pilot project Walmart rolled out in a Massachusetts store.[19] The protest gained enough traction that Walmart responded by saying it would not deploy the smart shelves. Protests then followed in 2005 and 2006 that focused on Tesco and Levi's use of RFID to tag products.[20] CASPIAN was a major player in all these protests, and Albrecht and Liz McIntyre (the communication director at CASPIAN) became a go-to source for journalists interested in quotes about RFID and privacy.

Protests against RFID were not confined to retail uses. In 2004, a coalition of thirty-nine privacy groups, including the American Civil Liberties Union (ACLU), the Electronic Frontier Foundation (EFF), and CASPIAN, collaborated on an open letter to oppose the use of RFID in biometric passports.[21] The protests did not stop RFID-enabled passports, but they were a precursor to later protests that focused on RFID in enhanced driver's licenses. In addition, concerns extended well past ID cards. People also began worrying about the mandatory implantation of RFID in human bodies. At least four US states—California, North Dakota, Oklahoma, and Wisconsin—passed legislation prohibiting the mandatory implantation of RFID chips.[22] And these concerns were far from exclusive to the United States. In 2006 in Great Britain, the *Daily Mail* published an article with the attention-grabbing headline, "Britons 'Could Be Microchipped Like Dogs in a Decade.'"[23]

Even with dramatic headlines, worldwide boycotts, and consumer protests, most people still knew little about RFID. In the mid-2000s, the little nonproprietary research that existed showed a minority of people even knew what RFID was.[24] A series of focus groups revealed low levels of knowledge and found that sentiments were generally negative. A 2004 nonrandomized survey of US adults showed that only 23 percent of adults were aware of RFID. As communication researchers James Katz and Ronald E. Rice summarized, "These studies show that public knowledge or even awareness of RFID technologies is low. Yet despite (or perhaps because of) this, fears about the technology's privacy threats are high."[25]

It was in this environment of low awareness about RFID—and especially low awareness of the technical capabilities of RFID—that one of the more important actors in the story entered the scene: the book *Spychips: How Major Corporations and Government Plan to Track Your Every Purchase and Watch Your Every Move*. The book—mentioned in chapter 5 because the authors published an alternative version targeted at evangelical Christians—was written by Katherine Albrecht and Liz McIntyre and is a dystopian account of what they label the "RFID menace."[26] *Spychips* is the single most comprehensive account of the privacy fears about RFID, and at one point rose to Amazon's list of best-selling nonfiction and was praised by mainstream outlets and publications such as the Associated Press and the *Chicago Tribune*.[27] Consequently, the book is worth discussing here in some detail.

Spychips is a difficult book to describe. On the one hand, it is an insightful piece of original research, and even the book's critics in the RFID industry recognized that the text brought up some valid concerns that had to be addressed.[28] Albrecht received her PhD from Harvard and devoted years to studying RFID technology through analysis of patent applications, interviews, and fieldwork. She and McIntyre made a compelling case that the real threat of RFID lay in a near-term future when tags and readers become ubiquitous. The picture the authors paint is straight out of Orwell's *1984* (referenced repeatedly in the text)[29] and imagines a world in which people's every movement is tracked by governments and corporations using RFID technology. The book's website even had a logo that read "RFID Nineteen Eighty-Four."

For all the value of the analysis throughout the book, *Spychips* sometimes comes off as overly inflammatory. The text is hyperbolic and dystopian, hypothesizing about what Stalin and Hitler could have done with RFID and forwarding governmental conspiracy theories about RFID and martial law.[30] Basically, most of the scenarios discussed in *Spychips* imagine the worst-case scenarios for RFID technology. To sum up *Spychips* in one sentence: "When RFID goes bad, it will be unlike anything we've ever seen before."[31] And, unsurprisingly, considering the rather polemical tone of the text, some of *Spychips* has not aged particularly well. In the decade since its publication, the worst-case scenarios have not come to pass. Industry players have somewhat addressed privacy concerns, and the public has likely become more accustomed to the technology as they see that RFID tags have not yet destroyed society. In addition, *Spychips* also sometimes exaggerated the technical capabilities of RFID, including the read ranges of some RFID tags. Although I cite *Spychips* throughout this chapter, I do so while tempering some of the book's more conspiratorial criticisms.

The actual peak of the public's concerns about RFID privacy concerns likely came about two years after the publication of *Spychips*. No definitive way to prove those dates exists, but sources support the claim. One source is a 2015 study by Daniel Castro and Alan McQuinn that analyzed Google search results featuring both "RFID" and "privacy" between 2000 and 2014.[32] Their research showed that searches about RFID and privacy peaked in the 2006–2007 period and began showing a downward trend in 2008–2009. A search for "RFID" on the Electronic Frontier Foundation's site that limits results to articles with the "privacy" tag reveals similar results, with 2005 and 2006 showing the largest number of articles. From another anecdotal perspective, the *RFID Journal* began publishing fewer articles about privacy in the late 2000s. The early years of the journal featured a recurring column filled with articles stressing that companies needed to avoid privacy controversies. As people stopped paying as much attention to RFID and the protests died down, the journal published fewer articles about RFID and privacy.

In one sense, the trajectory of privacy concerns about RFID followed a fairly typical pattern. The technology was initially met with fear, just like the telephone or the internet of the 1990s.[33] Then those fears lessened over time, even as RFID became more widely adopted. The reason people seem to pay less attention to RFID now, even as RFID is becoming much more ubiquitous, is impossible to identify with certainty, though a few explanations seem likely. For one, the overall privacy landscape is different in 2019 than it was in 2005. People carry phones that transmit their location multiple times a minute, they share pieces of their lives on multiple social media platforms, they make credit card purchases that are recorded, and they are tracked everywhere they go online. In addition, some forms of physical tracking, such as facial recognition used in supermarkets, make some of the examples discussed later seem comparatively tame.[34] Finally, as discussed throughout this chapter, many of the darkest predictions by privacy advocates (especially in *Spychips*) have simply not materialized.

Of course, the privacy concerns about RFID did not just disappear. The EFF has continued to warn people about the privacy implications of the technology,[35] the European Commission privacy working group published privacy guidelines companies should use when evaluating RFID,[36] and the American Civil Liberties Union has published more recent reports on RFID data abuses.[37] In addition, chapters 4 and 5 examined how RFID has remained a major target for certain groups, including some far-right online communities and groups of evangelical Christians. To look at the existing concerns, I next address individual use-cases of RFID, including in transportation, retail, workplaces, schools, passports, and in bodies.

Breaking Down Concerns

One of the challenges of writing a chapter on RFID and privacy was deciding where to focus. As discussed in the earlier chapters, RFID is a promiscuous technology that has found its way into everything from credit cards to smartphones to passports to library books. Without any exaggeration, it would be possible to write an entire book about potential privacy abuses of RFID technology (*Spychips* already did it). Rather than look at every possible issue, the following analysis instead focuses on a few related but distinct areas: (1) RFID and transportation, (2) item-level tagging, (3) workplace monitoring, (4) school monitoring, (5) biometric identification, and (6) bodily implantation.

To keep the discussion manageable, I examine issues directly related to RFID rather than broader questions about the ethics and privacy concerns of big data or the Internet of Things in general. Finally, after breaking down concerns, I conclude this chapter by situating my analysis of RFID within broader discussions about the contemporary state of privacy and looking more broadly at the ways granular practices of identification have already begun to reshape relationships among bodies and data.

Bodies and Transportation

In 2007, I spent a summer as an intern at a nonprofit environmental group in Washington, DC. The long commute for the job required me to ride the DC Metro every morning, so I bought a contactless SmarTrip card. The first time I walked through the Metro turnstile by waving my wallet at the reader seemed like magic. At the time, I did not think about how the system worked—it works through an RFID antenna and a microchip embedded in the card—or the implications of such a system.

The DC Metro is far from unique in relying on RFID-enabled infrastructure. As mentioned in chapter 2, the two earliest public transportation examples were the Korean Upass system (introduced in 1996) and Hong Kong's Octopus card (introduced in 1997). Similar systems now exist in cities in every inhabited continent, and one of the major implications of these systems is the ability to monitor data about ridership. Most RFID-based public transportation systems have someone use a credit card to load money on the "smart card." The card often has a microchip that stores the data, and the system tracks where people enter the system and deducts money accordingly. The contactless capability of the card does add convenience, but it also means people's movements are tracked throughout the public transportation terminals of the city. With paper tickets, people pay and then throw the ticket away. With the smart card, their card is connected to the individual purchasing information. Ostensibly, the location and exact times I entered and exited the system were logged in a database somewhere and attached to my payment information.

A Freedom of Information Act request submitted to Transport for London (TFL) provides a window into the types of data collected through public transportation cards.[38] In this particular request, an individual sent TFL the following message:

> My request is to know what exact information is stored on oyster servers about each individual card/person.
>
> EG do you have times/locations/costs/oyster card reader serial number/oyster card reader uptime etc stored on servers and linked to cards/persons.

TFL responded to the request for information by explaining,

> We retain full details of every transaction made by every Oyster card for a period of eight weeks. Full details include: card ID, date, time, location and device used (e.g. gate/ticket machine).
>
> We always know certain card-specific data, such as the date and place of issue, the date it was last used (though not where it was last used); the contact details for registered cards; and a history of the products (e.g. Travelcards) which have been loaded onto the card. We also retain details of any incomplete journeys made; journeys missing either the entry or the exit validation, which can be used for refunding cards. Finally, we keep some high-level data on the types of travel modes used (e.g. bus, London Underground, London Overground etc) which helps us to target our customer communications, for example when we know certain modes will be disrupted.

An FOIA request to the Washington, DC, transportation authority yielded similar results, with records showing the system collected detailed data about individual ridership.[39] Based on these existing records, one can assume that most, if not all, RFID-enabled public transportation systems collect individualized mobility data and store that data for variable periods of time. Consequently, by turning individual bodies into analyzable data points, they provide an example of the "liquid surveillance" mentioned in the introduction. As David Lyon wrote, "The concept of liquid surveillance captures the reduction of the body to data."[40]

As already established, RFID systems are used to monitor roads as well. Many toll roads all over the world enable people to pass through more quickly if they use an RFID tag attached to their windshield to log payment information. For example, states in the eastern and midwestern United States use E-ZPass, the Dallas-Fort Worth metro areas uses TollTag, Australian cities use E-Toll, Japan uses ETC, and so on. Just as with public transportation smart cards, these systems use RFID to communicate account information with toll readers, and that information is tied to an individual payment account. These systems are different from cash tolls, which do not generally keep records of people's movement.

RFID toll systems have been around since the 1980s and have achieved widespread acceptance. However, they have been criticized by privacy advocates because they create logs of where people go and the times they pass through tolls.[41] Figure 6.1 shows a small-scale example of the type of data collected. The image is a screenshot from my toll tag account that displays my automobility down to the minute. Because many of the Dallas-Fort Worth tolls enable people to pass through while traveling the speed limit, drivers may not even be aware their information is logged. In effect, cities are able to use the RFID toll system to build comprehensive records of how many cars there are on the road and the travel times between tolls.[42]

Tollbooths are one thing. People order RFID toll tags to facilitate movement and ostensibly choose to enter into the system, though the idea of choice is complicated because many systems charge much higher rates or do not allow access to certain roads

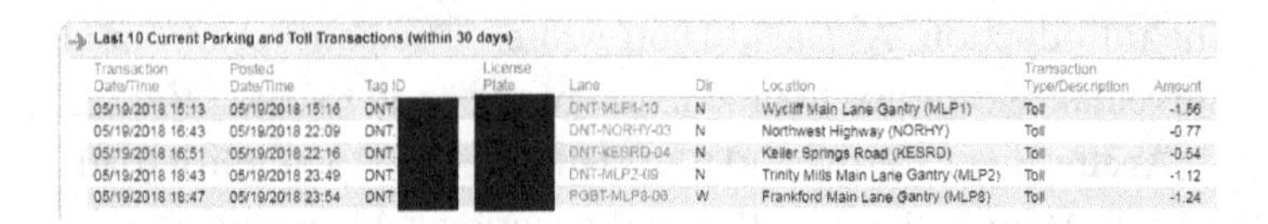

Last 10 Current Parking and Toll Transactions (within 30 days)

Transaction Date/Time	Posted Date/Time	Tag ID	License Plate	Lane	Dir	Location	Transaction Type/Description	Amount
05/19/2018 15:13	05/19/2018 15:16	DNT		DNT-MLP1-10	N	Wycliff Main Lane Gantry (MLP1)	Toll	-1.56
05/19/2018 16:43	05/19/2018 22:09	DNT		DNT-NORHY-03	N	Northwest Highway (NORHY)	Toll	-0.77
05/19/2018 16:51	05/19/2018 22:16	DNT		DNT-KESRD-04	N	Keller Springs Road (KESRD)	Toll	-0.54
05/19/2018 18:43	05/19/2018 23:49	DNT		DNT-MLP2-09	N	Trinity Mills Main Lane Gantry (MLP2)	Toll	-1.12
05/19/2018 18:47	05/19/2018 23:54	DNT		PGBT-MLP8-06	W	Frankford Main Lane Gantry (MLP8)	Toll	-1.24

Figure 6.1
A screenshot of my three most recent toll tag reads. Image courtesy of the author.

to people who choose not to participate. What is more alarming is that these tags can be read without people passing through marked toll readers. In 2013, an employee at the New York Civil Liberties Union (NYCLU) accompanied a privacy activist on a ride through New York City.[43] The privacy activist had created a device that signaled the driver whenever her E-ZPass was read. The device showed the car's tag was pinged over and over again without passing through any officially labeled toll system. The NYCLU then filed a FOIA request "to find out why E-ZPass readers had been installed far from any tollbooths for some purpose unrelated to their intended use, and what privacy protections, if any, were in place."[44]

The NYCLU found that the government had set up E-ZPass readers all over the state, and the readers were particularly prevalent in Midtown Manhattan (149 readers had been installed throughout the city by mid-2014). These readers logged the movement of individual vehicles. The data was used as part of a traffic management program, but people were not informed their data was being collected for these purposes. In addition, "the New York City Department of Transportation stated in response to the FOIL [Freedom of Information Law] that it had no policies or training materials on storage, retention, destruction or use of the information collected from its E-ZPass readers."[45] In other words, no plans were in place to govern the use and handling of the data. Instead, the city haphazardly collected people's mobility information without their knowledge, once again showing the rather invisible nature of RFID's role in object communication.

The New York City toll tag example is a reminder of how passive RFID technology works: it basically sits in an "off" state until activated by a reader. People carrying the tags do nothing to activate the tags except move close enough to the reader. They also do not receive an alert when the tag is activated. The data is collected from the tag and associated through a serial number with individual bodies linked to the account. The RFID tag in these cases works as an infrastructure of identification, but it also shows how infrastructure is always relational. On the one hand, the readers and the tags are an infrastructure of their own, but they represent the tip of the iceberg of a much larger infrastructure of back-end databases and various analytic techniques. As Karen Louise Smith and colleagues put it, in their discussion of RFID-enabled ID cards, the tags and readers themselves "provide very little sense of the massive infrastructure beneath the surface that is necessary for their functioning."[46] It is the hidden infrastructure that stores what researchers Ellen Balka and Susan Leigh Star call the traces of "shadow bodies," defined as "the reflections, illuminations and impressions of a body created through the illumination of some (but not all) of its processes, as captured in information systems."[47]

The concept of shadow bodies examines how individual data traces reveal small windows into the daily lives of bodies. The data produced through these RFID systems capture fleeting moments of movement, moments of passage through set points built with readers as identification infrastructure. While shadow bodies remain partial and fragmented, they can be aggregated into a fairly detailed picture of individual movement through an urban area. Where someone enters and exits transportation infrastructure provides the kind of surveillance data that can be paired with other data flows to provide more and more complete shadows of physical bodies.

Ultimately, just as with so many other examples analyzed throughout this book, each data point itself may seem unremarkable. But when physical infrastructure is built on practices of identification, with readers associated with specific locations, the data points begin to build over time. The shadow body, which always remains fragmented among diverse collection systems, nevertheless becomes more complete with each RFID read. The physical space becomes standardized and segmented through the placement of readers, and the tags themselves become traces of movement that can be compiled into comprehensive pictures of corporeal journeys through a physical environment. But transportation is merely one piece of the shadow body produced through RFID technologies.

Item-Level Tagging

One of the main surveillance concerns in *Spychips* focused on dense networks of readers tracking people's movement through the physical world. Central to these concerns is the growth of item-level tagging in retail. Item-level tagging was covered in earlier chapters, but as a reminder, it involves placing RFID tags on nearly all individual items to enhance inventory visibility. When the idea of the Internet of Things was first introduced by Kevin Ashton, he was describing item-level tagging.[48]

Spychips provides an in-depth account of the surveillance potential of item-level tagging, and the book goes into detail on various schematics companies used to describe potential RFID surveillance systems in the early 2000s. One example was the Gillette smart shelf. The "smart shelf" was designed to reduce theft and involved building RFID readers into retail shelving for Gillette razors (a commonly stolen item). The readers then detected when someone picked up the razors because the packaging included an RFID tag. Once the shelf detected tag movement, it used hidden cameras to take a picture of the person removing the razors. The picture would then be matched to people at checkout and used to deter theft. In 2003, Walmart canceled a trial to deploy Gillette's smart shelf after protests from CASPIAN and other privacy groups.[49]

The Gillette smart shelf involved using RFID in an isolated part of a store. Another patent *Spychips* focused on—"Identification and Tracking of Persons Using RFID in Store Environments"—was filed by IBM employees and described a much more comprehensive tracking scheme. Below is the abstract filed with the United States Patent and Trademark Office:

> A method and system for identifying and tracking persons using RFID-tagged items carried on the persons. Previous purchase records for each person who shops at a retail store are collected by POS terminals and stored in a transaction database. When a person carrying or wearing items having RFID tags enters the store or other designated area, a RFID tag scanner located therein scans the RFID tags on that person and reads the RFID tag information. The RFID tag information collected from the person is correlated with transaction records stored in the transaction database according to known correlation algorithms. Based on the results of the correlation, the exact identity of the person or certain characteristics about the person can be determined. This information is used to monitor the movement of the person through the store or other areas.[50]

The patent describes a full-fledged in-store surveillance system that could be used to enhance targeted advertising. In the design described above, items with RFID tags would not include personally identifiable information. But the system did not need to include identifiable information on tags to track people. Instead, to quote the patent, "If the collection of RFID tag information contains unique serial numbers identifying the products carried by the person, the likelihood of determining the exact identity of the person will be extremely high."[51] The system would use credit cards to obtain "personal information of the purchaser, such as the name, address, customer number, etc." and associate that information with RFID tags that are "integrated into items" and are "extremely small and inconspicuous."[52] What is possibly most revealing about the patent is a throwaway line midway through the application: "The system will be explained for use in a retail store, but is equally applicable in other locations or geographical areas."[53]

The system described in IBM's patent was never actually created (to the best of my knowledge), but it shows why RFID concerned privacy advocates so much. What IBM's patent and a few similar early-2000s patents essentially imagine is a future in which people's behaviors are tracked in the physical world like they are online.[54] Online, people's movements are already tracked and their demographic and browsing information is used to deliver targeted advertising.[55] IBM's engineers imagined bringing that model out into the physical world. Online, if someone went to a hardware store website to look for a lawnmower, they would probably later see an ad for a lawnmower. Targeted advertising through RFID tracking would possibly work the same way. When someone picks up an item, they then would receive ads for similar items when they pass by some kind of advertising module.

But alas, fifteen years after the patent was submitted, no one has created a system on the scale described in the IBM patent.

One reason might be that the protests that rocked the RFID industry in the mid-2000s halted such plans and forced companies and industry leaders to acknowledge privacy issues. Also, some of the impetus behind RFID tracking for advertising has been lessened because of the growth of location-based advertising linked to mobile phones. Mobile phones already provide comprehensive logs of people's locations, and some of that data is used to target advertisements. Building a wide-scale network of RFID readers seems more complicated than buying into a system like Verizon's Precision Market Insights to target customers.

However, the idea of tracking people through stores has certainly not disappeared, especially with the recent return to the idea of the cashier-less stores like Amazon Go. Amazon Go, which was opened to the public with much hype in 2018, does not use RFID and instead uses cameras and facial recognition. Other cashier-less stores, such as the BingoBox stores in China, do rely on RFID to power their systems. For those stores to become a more widespread reality, the space will have to turn into a site of near-constant surveillance. Each item will have to be uniquely identifiable and associated with individual bodies. The bodies will have to be associated with accounts that can be charged based on the item reads. The stores, whether through RFID technologies or cameras and facial/item recognition, as in the Amazon Go store, will be spaces of powerful automated identification and differentiation. The separation of similarity I discussed in chapter 4 will power these stores through various infrastructures of identification, and identification of bodies requires surveillance. The limits placed on what can be read and what data can be collected and stored will be important to follow as labor practices continue to shift toward the further automation of retail.

So how serious a privacy concern is RFID in the context of transportation and retail? Well, the examples discussed above provide reason to pay attention. The New York City example shows how comprehensive an RFID tracking system can be. The massive number of people sharing mobility data on public transportation shows how ubiquitous mobility data can be. The IBM patent shows how retail stores could build RFID systems to further surveille customers, and the patent possibly predicts data practices that may shape future automated stores. But it is important to note that, at least for retail, the vast majority of RFID item-level uses are nothing like the IBM or Gillette examples. Most stores that use item-level tagging do so for supply-chain logistics, in-store inventory, and new forms of retailing, not for the surveillance of individual consumers. In fact, even as item-level tagging has begun to possibly approach a supposed "tipping point," no prominent recent stories on the level of the mid-2000 CASPIAN protests have roiled the RFID industry. However, as I discuss at the end this chapter, simply the potential for widespread surveillance, whether by governments or corporations, shows the importance of embracing a sense of the "infrastructural imagination."

Workplace Privacy and RFID

One of the most unique uses of RFID I came across is an RFID-based handwashing system found in some hospitals.[56] These systems include an RFID reader installed above a sink that detects the RFID bracelets employees already use for access and for being located in emergencies. The bathroom monitors how long people's bracelets are within range of the reader; the RFID system will tell hospital administrators if someone washed their hands for only five seconds instead of the minimum twenty seconds recommended by the US Centers for Disease Control. Many hospitals in the United States deployed the system because the spread of communicable disease is a problem in health-care facilities, in part because employees sometimes do not follow proper hygiene procedures. A follow-up study in an Alabama hospital showed the system contributed to a 217 percent improvement in handwashing compliance, and at least five different companies make their own RFID hygiene systems.[57]

On the one hand, the benefits of such a system are clear. Patients want to know that members of the hospital staff have washed their hands properly. On the other hand, the hygiene monitoring system shows the workplace surveillance potential of RFID.[58] Employees all over the world are already monitored by RFID technology, whether they know it or not.[59] For example, many jobs require people to wear badges to access certain areas. Some employees have key fobs they use instead of traditional keys to enter buildings. These relatively mundane examples of workplace technology are typically RFID systems, and they produce data about employees' movements. Access badges can tell an employer how long someone spends in an area of an office, when someone leaves work, or even how long someone accesses a record on a computer.[60] Some construction companies use RFID badges to monitor which employees use pieces of equipment.[61] In 2004, Mexico's attorney general and at least 160 of his employees received RFID bodily implants that monitored access to certain secure buildings and documents.[62]

Here it helps to give an example of how RFID data can be used without employee knowledge. In July 2003, a man crashed into a busy California farmer's market. The accident killed ten people and injured many more. The farmer's market was located near one of RAND Corporation's offices, and the accident occurred around lunchtime. RAND employees were worried coworkers might have been at the market at the time of the accident, and "using data collected through RFID employee access cards, managers were able to determine who had left the building before noon but had not returned. Then, they were able to go to those people's departments and find out why they were absent."[63]

No RAND personnel were hurt in the accident, but the office visits came as a shock to the employees. They did not know their RFID badge data could be used to track them, and when employees began looking into policies regarding employee badge data, they found the company had no policies in place governing who could access the data or what could be done with it. And this situation was not unique. Research on workplace privacy shows that companies increasingly track people's movements and that employees often have little power over or knowledge of how their data is handled.[64] As the RAND example suggests, waving a badge in front of a reader can become so mundane that people do not realize how much that data reveals.

Even if employees are made aware of RFID tracking, it is unlikely they are given much of a choice about whether or not to participate. They will likely be interpellated into various surveillance networks as a requirement for their job, especially considering that some more extensive forms of tracking are introduced using rhetorics of safety and security. An illustrative example is Lettire Construction's use of Spot-r RFID technology. Spot-r involves active RFID and sensors located in a work belt worn by employees at construction sites. The system provides supposed "total digital worksite visibility," including the real-time location of employees and data about who accesses specific parts of the site. As the following quote from the CEO of the company that produces Spot-r shows, the data visibility is presented as a benefit for the individual employee: "By incorporating Spot-r, Lettire site supervisors and company management—no matter where they are located—can now see where workers are on the jobsite and receive alerts to any safety issues in real-time. This level of visibility and actionable data helps them proactively improve safety, protect their workers, and achieve greater productivity."[65] I have no reason to doubt that technologies like Spot-r can improve workplace safety, but they also further tilt the already slanted power dynamic between employee and employer. With the Spot-r example, employees can possibly have every minute of their workday tracked through RFID location data. And they likely have little choice in the matter.

Of course, like with almost any form of surveillance, the scope and permissibility of employee surveillance through RFID depends on cultural context.[66] An example of different levels of legal and social permissibility can be found in an insightful dissertation by health informatics researcher Chandana Unnithan.[67] Her dissertation analyzed the adoption of RFID to track objects and monitor pharmaceuticals in two Australian hospitals. As she writes, "Privacy and legal issues within the Australian health context impedes people tracking to a significant extent."[68] Later in her dissertation she discusses how, "if surgeons felt that their productivity was being tracked, that would raise privacy issues, and unions would also enter the fray,"[69] and she found similar issues with the tracking of nurses. In sum, Unnithan's work showed how, at least in Australia, hospital workers had relatively robust legal frameworks and significant union support that prohibited tracking them through RFID in the workplace.

Her findings in Australian hospitals stand in contrast to research by social medicine researchers Jill Fisher and Torin Monahan in hospitals across the United States.[70] In a wide-ranging study, the researchers tracked the spread of RFID through the US health-care system and reported on how RFID was used to surveille patients and hospital workers. Regarding RFID, the researchers found that nurses expressed "concern that they are overly scrutinized by these surveillance technologies."[71] Going further, "nurses describe feeling like 'big brother' is watching as they spend time with patients, take unofficial rest breaks between patients, and take official breaks during their shifts."[72] Fisher and Monahan's work is an example of how even the threat of surveillance can exert a chilling effect on behavior. Importantly for this discussion, their work also shows how different privacy law is in the United States and Australia. Many of the practices Fisher and Monahan covered—for example, the tracking of hospital staff—were prohibited in Australian hospitals for legal and cultural reasons.[73]

A full account of individual workplace privacy laws is outside the scope of this chapter, but workplace monitoring through RFID has already become widespread, and anyone who uses an RFID-enabled badge to enter an office can be monitored by their employer. Once again, these examples show how RFID technologies can use identification data to turn previously ephemeral practices into permanent data traces. The data may follow a construction worker who steps outside to take a smoke break, a nurse who lingers a few minutes too long in the break room, a physician's assistant who does not put a scalpel back, or an office worker who uses her badge to leave the office a few minutes early. The RFID tags are associated with each of those bodies, and the unique identification enabled through the technology plays a role in sorting and tracking behaviors across large spaces standardized through the positioning of readers. As I have argued throughout this book, RFID has the potential to turn almost anything into a digital trace, and that is just as true for the workplace as it is for the global supply chain.

School Monitoring

RFID bracelets and badges have been used to automate attendance and monitor schoolchildren since 2004. In Brazil for example, more than twenty thousand students wear "intelligent uniforms" that feature RFID chips used to take attendance and track the students.[74] These systems often work in a way similar to the employee monitoring described above: children either wear clothing with RFID or wear an external object like a bracelet or necklace that contains an RFID tag. The systems have sometimes been successful and sometimes highly controversial. The difference between the two poles, just as with most of the other examples discussed throughout this chapter, often has more to do with the social than the technological, more to do with the discourses surrounding the school RFID system than the actual system.

Generally, school RFID systems have been more successful when schools inform parents of the benefits of the system and provide the option of opting out. Two contrasting examples can show why.[75] Possibly the first school RFID system was deployed in Spring Independent school district near Houston, Texas, in 2004. The system was deemed successful and eventually expanded, in no small part because school administrators worked to give parents choice and inform them of the safety benefits of RFID for their children. On the other hand, Brittan Elementary School in California rolled out an RFID project (funded by a company called Incom so they could test their product) and were heavily criticized.[76] In the Brittan case, school administrators did not inform parents. Children were just given

RFID badges to monitor their behavior, and the project did not become public until parents complained. Unsurprisingly, parents and the ACLU stepped in to protest, and Brittan Elementary canceled its RFID plans.[77]

These are two isolated examples among many cases, both successful and unsuccessful. In the United Kingdom, a high school in Doncaster began a project that involved RFID chips sewn into blazers that monitored student location, but the trial was stopped in part because of privacy concerns.[78] West Cheshire College went a step further than any of these examples, asking students to wear active tags around their neck (compared to the passive tags used elsewhere). These tags transmitted student location every second and enacted what I described in earlier chapters as a real-time locating system (RTLS) for students.[79] The West Cheshire project was cancelled in 2013 after three years. Similar RTLS systems have also been used in schools in Germany and the United States.[80] In a notable 2013 case related to the "Mark of The Beast" discussed in chapter 5, an evangelical Christian student in Texas was suspended for refusing to wear an RFID-enabled badge to track her movements.[81]

The monitoring of school-age children with RFID brings up larger issues about children's privacy. As lawyers Benjamin Shmueli and Ayelet Blecher-Prigat wrote in an article for the *Columbia Human Rights Law Review*, the "literature suggests that today's children are 'the most watched over generation in memory.'"[82] The push toward increased monitoring of children is often framed with the "language of safety, protection and care" and "monitoring has become associated with good parenting."[83] RFID systems are framed similarly, and in some cases, such as a deployment in Germany, are introduced as a specific response to a school shooting. Consequently, it is unsurprising that some parents are fine with a school monitoring their children through RFID, at least as long as the parents are informed about the system.

However, there are deeper concerns about children's privacy even when parents are informed. For one, our current culture of monitoring has been criticized for inhibiting youth behavior and normalizing young people's experience of ubiquitous surveillance.[84] Children are taught from a young age to expect to be watched constantly, and as criminologist Emmeline Taylor points out, schools are often the testing ground through which practices of surveillance are made commonplace.[85] More specific to RFID in schools, privacy advocates have identified potential ways comprehensive RFID data could be abused by school administrators. Some of these concerns seem unlikely, but they are technically feasible and worth keeping in mind when discussing school uses of RFID. The following paragraph from an Electronic Frontier Foundation article lays out this potential dark future:

> If readings are taken often enough, you could create an extraordinarily detailed portrait of a child's school day—one that's easy to imagine being misused, particularly as the chips substitute for direct adult monitoring and judgment. If RFID records show a child moving around a lot, could she be tagged as hyper-active? If he doesn't move around a lot, could he get a reputation for laziness? How long will this data and the conclusions rightly or wrongly drawn from it be stored in these children's school records?[86]

The quote above encapsulates many of the concerns about RFID: people's fears often focus on how it may be used in the future. The same fears apply to big data in general. Just the fact that so much data exists about various phenomena means the data can be abused. RFID is no different. That possibility does not necessarily mean that RFID should not be used in schools under any circumstances. Banning a new technology because of potential misuses is a slippery slope that would likely result in a freeze in innovation. However, identifying potential abuses is an important way to positively shape innovation. Being aware of what RFID does, what data it transmits, and how that data can be used is vital to creating school systems that do not abuse student privacy and do recognize the limits of data-informed decision making.

Biometric Identification

I renewed my US passport in 2014 and received a new document about a month later. The passport was slightly different aesthetically, but it contained all the same information on the main page. The new passport featured one major difference, however: RFID (see figure 6.2). Beginning in 2006, all new US passports include an RFID chip that contains biometric information, a move predated by eight years by Malaysia's rollout of the world's first ePassport.[87] European Union and Chinese passports also now include RFID chips that include biometric information.[88] As of 2016, more than fifty countries located on six continents require RFID-enabled biometric passports, and in 2015, the US Department of Homeland Security passed new regulations stating that anyone traveling to the United States without a visa must possess a biometric passport.[89] And RFID is not exclusive to passports. Some enhanced driver's licenses in the United States and enhanced national ID cards in countries like Finland, China, and Portugal also include RFID chips.[90]

Biometric passports were widely adopted as part of increased security measures post-9/11. They are designed to provide machine-readable information about the passport holder to make the passport more secure. However, they have been criticized for much the same reason as contactless payment cards: they

Figure 6.2
Passports pre- and post-inclusion of RFID.
On the left is an old, pre-RFID passport. The two other passports contain the international symbol for RFID-enabled passports. Photo courtesy of the author.

can be skimmed by hackers.[91] The RFID chip in passports (and other enhanced IDs) contains a wealth of sensitive information, potentially including pictures, fingerprints, date of birth, and place of birth. These concerns are why a coalition of privacy groups wrote an open letter in 2004 protesting the use of RFID in biometric passports.[92]

To understand why some advocates have complained about ePassports, it is first important to understand how biometric passports work. They rely on the ISO 14443 standard, which requires use of an HF RFID chip.[93] The frequency means the chip is intended to be read from around ten centimeters away, though some trials have reportedly shown the passports can be read at longer ranges depending on reader technology and environmental factors.[94] And while different biometric passports use different levels of encryption, security experts have shown that most ePassports have security risks and can be skimmed by dedicated hackers.[95] Nonetheless, RFID-enabled passports are now required by many countries as a way to supposedly respond to the growing risk of terrorism and control flows of mobility across national borders. They have been pushed through using a rhetoric of risk and security that fits within the larger discourses of security that shape contemporary risk society.[96]

In their defense, governments have taken steps to increase passport security. Many passports have RFID-blocking covers that keep chips from being read unless the passport is open. Newer ePassports now often use "basic access controls"—first used in German passports in 2005 and added to US passports in 2007—that ideally require physical access to the passport to create a key to control access between a reader and the passport chip, though critics claim the key can be guessed through brute force attacks.[97] In addition, some of the fears about passport skimming are possibly overstated. A study by a group of Canadian surveillance scholars showed how. The study did not focus on passports, but rather on the RFID included in enhanced driver's licenses in Canada.[98] The project involved a playful intervention in which people were handed a card using RFID technology identical to that in enhanced driver's licenses and then had their data scanned from a distance and displayed on a screen. When the researchers designed the intervention, they expected to be able to scan the cards from significant distances because of read ranges that had been found in laboratory settings. However, "laboratory tests under ideal conditions show a read range of up to 50 metres but in the 'real world' setting of live public events, subjects appear to have to willingly collaborate with the researchers to produce any results."[99] Their findings are also relevant to the discussion of biometric passports. Although passports can conceivably be read from longer distances, in practice that remains unlikely. The actual ability to read passports is limited by a variety of factors, and a hacker would have to bypass those factors and then also break the basic access controls to read the biometric data.

Security, however, is only one issue with the use of biometric passports. Even if these passports are secure, it does not mean they are not questionable technologies of surveillance. Each RFID chip links to biometric information. Biometric information is itself an infrastructure of identification used to sort bodies. In fact, biometrics are one of the most prominent data points that feeds into the type of "social sorting" David Lyon identified as one of the key social consequences of surveillance.[100] RFID, in the case of passports or ID cards, is an infrastructure that mediates a larger infrastructure of sorting and differentiation linked to extensive databases containing the physical characteristics of human bodies. Few consequences of social sorting are more significant than who gets to cross a national border. Consequently, the use of biometric information, such as digitized facial recognition and fingerprinting, to control and monitor transnational mobility has been criticized repeatedly from a surveillance perspective.[101] A full account of the issues surrounding biometrics is outside the scope of this chapter, but RFID's rather hidden place in that debate once again shows the role the mobile technology plays as a key infrastructure of identification.

Bodily Implantation

In October 2017, Mike Miller, the chief executive of the World Olympians Association, floated the idea of microchipping Olympic athletes to monitor them for doping. He was quoted as saying, "Some people say we shouldn't do this to people.... Well, we're a nation of dog lovers, we're prepared to chip our dogs and it doesn't seem to harm them, so why aren't we prepared to chip ourselves?"[102] While it is impossible to know exactly what type of chip he was talking about, it was likely some form of an HF RFID chip paired with biosensors. Unsurprisingly, his idea was met with immediate outrage and led Liz McIntyre, one of the coauthors of *Spychips*, to create a new advocacy group: Citizens Against Marking, Chipping and Tracking (CAMCAT).[103] The larger issue of RFID implants and privacy has also been examined in an edited collection called *Uberveillance and the Social Implications of Microchip Implants*,[104] which provides an extensive discussion of the possible implications of RFID microchipping. In chapters 4 and 5 I looked at how subcutaneous RFID is a technology that elicits particularly strong responses from some groups; here I want to expand on that discussion. But I also do not devote too much space to implants and surveillance, in part because I believe we are still a long way off from any widespread implantation and in part because I already analyzed conspiracies about implants and fears from evangelical communities.

RFID implantation has been proposed to combat identity theft, facilitate health care, replace keys, and monitor athletes. Some patients, especially elderly patients and mentally disabled patients, have received implants to track them for medical purposes, though the practice is still far from widespread.[105] While one small sample size survey did find a slight increase in the percentage of college students willing to implant RFID,[106] Katz and Rice found that a national survey "suggests that while there is not likely to be widespread antipathy for topically applied devices, the public is somewhat less accepting of subcutaneously inserted devices."[107] Or, as Monahan and Fisher put it, "Unlike most emergent technologies, they [RFID implants] are perceived as a fraught device, with negative potentialities that must be discussed before widespread adoption."[108] Consequently, it remains fairly unlikely that widespread RFID implantation will happen in the next decade.

In addition, while the surveillance and privacy concerns have been the main focus of scholarly discussions of RFID implantation,[109] one of the few empirical studies on the topic found different ethical concerns than predicted in the theoretical literature. Monahan and Fisher, in a report on interviews with workers in the US health-care industry, wrote, "While we are entirely sympathetic to and in agreement with most of the recommendations made by bioethicists, information ethicists, and privacy advocates about these devices, they are not grounded in the empirical realities of US hospitals."[110] The three major ethical concerns they found—the furthering of social inequality, the privileging of chip information over patient input, and the endangerment of patients through lack of informed consent—were more prevalent than surveillance concerns. As the authors note, their findings do not mean people should not be concerned with the surveillance potential of implanted RFID; rather, they show that the discussions of surveillance that dominate literature on implantation do not accurately reflect the "on the ground concerns" and may obfuscate less obvious ethical issues associated with RFID implantation.

And as a final note, it is important to remember the technological limitations of RFID: small HF chips have short broadcasting ranges. UHF RFID, which has a longer read range, is limited by water, and human bodies are mostly water. So for a chip to broadcast long distances, it would likely have to be much bulkier than the chips implanted in pets and the HF chips found in most biohackers. Finally, and possibly most importantly, passive RFID chips are not GPS transponders. They are not always-on forms of locative media; rather, they transmit only when near a reader. That point sounds obvious, but research on public sentiment toward RFID injections has found that "the biggest concern of the respondents was the possibility of GPS tracking of microchips."[111] The future may be the cyborg's dream and the privacy advocate's nightmare, and the issue is one worth watching. However, we are still likely a long way off from any sort of widespread implantation. If that day ever does come, people should be made aware about what microchips can and cannot do.

Bodies, Data, and Metaphors of Privacy

One popular way people have understood the surveillance potential of RFID is through the metaphorical lens of George Orwell's Big Brother. The metaphor imagines an all-seeing entity monitoring the population, and *Spychips* often turned to that metaphor, as have discussions of the huge database of biometric mobility data built through RFID-enabled passports.[112] In academic circles, especially in some wings of cultural studies and surveillance studies, people may prefer the Foucauldian metaphor of the Panopticon, in which a central observer can watch the populace.[113] These two metaphors often dominate debates about privacy.[114]

I argue here, however, that neither the Panopticon nor Big Brother accurately capture the privacy implications of RFID, and both metaphors can obscure more than they reveal. After all, RFID is found in diverse places and produces identification data that can be accessed by a diverse set of actors: toll data and subway card data is often stored by cities, contactless payment data stored by banks, passport data stored by national governments, inventory data stored by retail stores, workplace data stored by employers, and so on. The account in this chapter does not point to an all-seeing Big Brother or a guard inside a Panopticon. Instead, the data produced about people's movements is potentially accessible to a wide range of actors, depending on the setting and individual use. I argue that the ubiquity that makes RFID an interesting infrastructure of identification also necessitates a different metaphorical approach for understanding potential privacy implications.

One place to find such an approach is the work of legal scholar Daniel Solove.[115] Solove has argued for the need to adopt a new metaphor for understanding the contemporary privacy landscape, and the metaphor he chose informs much of my understanding of privacy in relation to RFID: Franz Kafka's *The Trial*. *The Trial* tells the story of a man who is arrested and never informed what he is being charged with. He is then released while awaiting trial, but he has no recourse to learn about the charges or see the evidence against him. When he does come in contact with his accusers, he faces an incompetent bureaucracy rather than an all-knowing watcher. In short, he has no idea what is going on.

Drawing from Solove's work, I argue that *The Trial* provides better conceptual tools for understanding RFID and privacy than the more dominant metaphors of the Panopticon or Big Brother. People have little access to how the data they produce through RFID is used, how frequently it is collected, or how it is stored. Going even further, many people likely have little idea they even come into contact with RFID technology when they pass through a terminal, drive through a toll, or buy a product from certain retailers. Consequently, the issue is not so much that people are being actively surveilled through those data sources (though they could be, in certain cases); the more pressing concern relates to the lack of knowledge and power. Additionally, much of the concern about privacy in these instances has just as much to do with the databases the RFID technology communicates with as it does with RFID itself, and the general public has no way of knowing how secure a database is.

Digital identity—the capturing of the "shadow body" or "digital double" or whatever phrase one prefers—is fractured and partial, with different databases capturing parts of a larger whole. The data covered in this chapter mostly focuses on identification, and "identification infrastructures seek to reliably link an individual to some unique identifier."[116] The unique identifiers enabled through RFID for surveillance purposes are multiple and varied. One person may transmit different identifiers on a daily basis when they work in an environment with RFID badges, pass through a subway terminal to get home, drive through a toll to get groceries, and then use an RFID-enabled card to pay for the groceries. The view is partial but nonetheless consequential and representative of the larger shifts in practices of identification that include a multitude of technologies besides RFID. And ultimately, the granular practices of identification represent the move covered in chapter 4 to sort similarity and reconfigure the idea of "mass" in contemporary society. Mass-produced items in an automated store are no longer the same if each is differentiated by an EPC; masses of bodies moving through a subway are no longer a mass in the traditional sense when each is linked to a specific account; the mass of students who enter a school each morning are identifiable when each has a bracelet with an antenna inside.

Importantly, the surveillance of most of these identification infrastructures focuses more on control than discipline, a concept I covered in chapter 5. The identifying information controls access to road infrastructure, public transportation, payment networks, border crossings, and so on. The data could be used at some point as a form of discipline, but the immediate consequences of these fractured forms of surveillance are much more about sorting and enabling behaviors than punishing immediate actions. And the systems we build increasingly rely on surveillance and identification to function. The RFID-only toll roads demand traces of identification data to be accessed; the workplace with RFID often will not provide a key that does not track access.

Ultimately, the varied and partial nature of the identification capabilities of RFID infrastructures is more similar to the messy uncertainty of *The Trial* than to the dark "RFID menace" discussed in books like *Spychips*. And metaphors do matter. I do not suggest *The Trial* as a framework for understanding RFID merely as an academic exercise. Metaphors shape how people understand the world,[117] and looking at RFID through a lens of uncertainty provides a different view than looking at RFID as a tool of total surveillance and discipline. However, that uncertainty in itself should raise concerns. Most countries have no legal frameworks governing the use of RFID, and at this point, people pretty much have to trust state and corporate actors to not abuse the technology. The New York example shows how one state developed an RFID surveillance system without people's knowledge and then seemed unprepared to protect that data. The IBM and Gillette patents show how companies have at least explored building large-scale in-store surveillance systems.

At this point, people have to trust others to act responsibly. To the best of my knowledge, companies and cities are not actively surveilling people through RFID. But the key part of that sentence is found in the introductory phrase: "to the best of my knowledge." The reason a metaphor like *The Trial* is so useful for understanding the privacy landscape of RFID is that people just do not know. I talked to more than thirty people in the course of researching this book; I read everything I could find on RFID, both in the popular press and in academic venues. But I still cannot say for sure what is going on. I might know better than most because I have searched out RFID infrastructure, but the fears about the ubiquity and invisibility of the technology are real; there is no real way to know for sure.

On the one hand, it can seem a bit conspiratorial to automatically assume companies, government, employers, and schools will put the data they collect to the most invasive uses. As one figure in the RFID industry told me, maybe companies should get the benefit of the doubt because, "even as item-level tagging has increased, there hasn't been a single major case in the last decade of companies using RFID to abuse people's privacy." In addition, some standard-bearers in the industry have attempted to address privacy concerns. GS1 introduced labeling to identify products tagged with EPC RFID (see figure 6.3). The standard also suggests a "kill" option so tags can be turned off once a product is purchased. In addition, companies that have adopted item-level tagging have taken some measures to protect privacy. American Apparel and Marks & Spencer, for example, include only RFID tags that are meant to be thrown away. ePassports have become more secure as countries have improved encryption and added RFID-blocking covers.

On the other hand, fears expressed about RFID make sense when understood within our broader information landscape. Take

Figure 6.3
The EPC "cube" used to identify RFID on products and packaging. Source: GS1 (https://www.gs1.org/standards/epc-rfid/guidelines).

the worries about corporate surveillance as an example. Retailers have seemed unwilling to risk privacy controversies by tracking customers in plans such as those laid out in the patents discussed earlier. However, consumers do not have to look far to find examples of companies engaging in fairly similar behaviors with other communication technologies. People's digital movements online are tracked every day by a variety of corporate actors. Communication researcher Joseph Turow's book *The Daily You* discusses online tracking in detail, showing how cookies track people across sites and help companies build targeted advertising profiles based on people's behaviors.[118] As Turow points out, most people have little idea of the extent to which their online behaviors are monitored. With RFID tags that can be sewn into clothing and put through the washing machine, one can see the potential—though it is mostly just potential at this point—of online tracking and monitoring moving into the physical world through RFID.

Or take the fears about transportation cards or electronic toll collection as another example. People who use these systems leave a digital trace of their mobility, often without thinking twice about it. They do not often know how comprehensive the records are, who has access to the data, or how it might be used. The New York City example shows how such systems can extend beyond initial intended uses. That example was used to measure traffic data, but each of those tags is linked to an individual's account, and the data could easily reveal sensitive information. Furthermore, there is little reason to think such systems would be beyond the power of state actors. As many of the documents leaked by Edward Snowden showed, the scope of state surveillance is broader than many people expected and often involved location data produced by mobile phones.[119] Using toll data or public transportation data as part of the national "surveillant assemblage" seems like a fairly logical step.

In addition, issues of power are important for understanding RFID and surveillance. The relationships described above represent unequal power dynamics. In some areas, people have the option of foregoing RFID toll technology, but they have to pay almost twice as much per toll if they use cash and may not be able to travel on all roads. In some cities, people can bypass contactless RFID transportation cards, but they have to be inconvenienced by slower systems. In retail, most people will not know item tags include RFID. If they follow CASPIAN's advice, consumers have the power to boycott retailers that tag items with RFID, but most people will not have the information to know whom to boycott. Many people also are not able to avoid shopping at certain stores, especially as RFID becomes more widespread. As research on credit cards or online browsing has repeatedly shown, people sacrifice privacy for convenience and often feel they have little choice in the matter.[120]

One answer to the issues explored in this chapter could be legislation specific to RFID. However, governments have shown little willingness to legislate RFID, even at the height of the privacy concerns of the mid-2000s. That hesitancy is fairly typical, considering RFID is still an emerging technology, and passing strict legislation targeting problems that often do not exist yet threatens to stall the technology before it has a chance to grow.[121] In addition, RFID is already embedded in various transportation infrastructures, so legislation at best could target specific practices of data collection. Consequently, although increased legislative limits on RFID data collection may be one way to approach issues laid out in this chapter, it is unlikely many governments will pass sweeping legislation anytime soon. Considering that, I conclude with some broad suggestions for how individuals can exert some agency in regard to privacy and RFID.

Conclusion

The surveillant assemblage can feel unavoidable. Short of clearing all browser cookies and using specialized search engines like DuckDuckGo, it is difficult to avoid being tracked online. Short of artistic interventions to map the lines of site of CCTV cameras, it is difficult to know when one's image is captured while mobile in the city. Short of literally not using a mobile phone, it is almost impossible to not share location data with service providers. And RFID is already difficult to avoid and will be even more so in the future. Unless one manages automobility through cash tolls, relies on paper tickets on public transportation, does not travel across passport-controlled borders, and keeps tabs on companies that use RFID at the item level, people will have their data shared through RFID technologies.

The situation above can seem overwhelming. However, people are not powerless in the face of new technologies. I want to conclude this chapter with a few suggestions for how people can exert some agency when it comes to privacy, particularly privacy in relation to RFID technology. One way to exert some control is through what I have been calling the "infrastructural imagination." One of the main goals of this book has been to peel back the surface layer of various objects to examine the underlying RFID technology that gives voice to things. Few people have the time to search store shelves for RFID tags, tear off the plastic on their toll tags or subway cards, or read about the ISO standards on ePassports. I recognize that. Nonetheless, the recognition of just how much nonhuman communication happens that is hidden from view is a good place to start when thinking about engaging with RFID and potential privacy issues. A willingness to think about infrastructure, to spend an extra thirty seconds reflecting on how that interaction with a badge and an office door actually works, is a strong place to start.

Of course, the infrastructural imagination involves more than identifying material objects in the environment. The practice also includes understanding the impacts of those objects. Consequently, acknowledging the practices of identification enabled through RFID is an important step in understanding the consequences of the technology. As I have argued throughout this book, infrastructures of identification potentially reshape the relationship between objects and bodies by making each object unique, even if that uniqueness is solely contained in RFID data. The sorting of similarity means that objects that may appear identical are not identical. One shirt is not the same as identical shirts as far as a retail database is concerned. One plastic wristband is not the same as another as far as an access system is concerned. Fundamentally, exploring RFID as infrastructure involves finding tags and thinking about the work those tags do to separate and differentiate objects.

Developing an infrastructural imagination is also one way to head off potential abuses. People can identify RFID tags on clothing and inquire about how the tags are used and whether they are deactivated at the point of sale. They can look for RFID readers located on store floors or RFID readers throughout the city that ping tags. They can ask how their children are monitored if a school installs an RFID system. They can inquire about how their employers use RFID access data. Many RFID readers will be difficult or sometimes impossible to find; many tags might be so small that they cannot be seen. My suggestions here and the examples I give in this book will not solve privacy concerns. After all, what makes RFID so interesting—its potential ubiquity, promiscuity, and small size—are also what make it so concerning to privacy advocates. Few people will have the time to track all RFID developments or identify every time they come in contact with the technology; however, learning how the systems work at a general level can at least help people prepare for and identify potential privacy issues associated with the technology.

The infrastructural imagination as applied to RFID can be a step in the direction of exerting some control over the data one shares about movement, consumption, or even biometric physical characteristics. Any exploration of RFID infrastructure, however, will be limited because of the relational nature of infrastructure. RFID is a built infrastructure that consists of tags and readers and middleware. People can exert some agency in exploring and understanding RFID technologies. But RFID also works as an infrastructure that supports much deeper levels of infrastructure that include huge databases, security procedures, and software for analysis. Almost no amount of exploration will give people access to the inner workings of the buried infrastructure that supports RFID as an infrastructure of identification. As someone who has spent the last few years researching RFID, I understand how different types of tags work. But no amount of exploration can tell me if the North Texas Transportation Authority databases are secure or if the DC Metro secures my data. Consequently, the infrastructural imagination as I explain it will almost always be partial. It can grant access to knowledge about the material layer of RFID data, but it will always struggle to reveal the adequacy or design of the underlying infrastructures that hosts the data. After all, as the seemingly steady stream of hacked services has shown, data is secure right up until it's not.

Despite its limits, the infrastructural imagination of RFID is still valuable. As our built environments increasingly rely on precise practices of identification, it is important to understand how differentiation happens. In many cases, one cannot understand that process without knowledge of RFID infrastructure. After all, the mass production of identification and mobility data has the potential to be abused. Workplaces could engage in deeper surveillance; schools could make decisions based on RFID-related data; cities could monitor automobility at unprecedented levels; people could someday have their movements tracked by RFID implanted in their body. All these examples are possible. At this point in the development of the Internet of Things, people face a moment of forking paths, and decisions must be made about what the public is comfortable with sharing and what kinds of data should not be collected and acted on. For people to make those decisions, they must learn about consequences, technical capabilities, and conveniences. This book is only one small step in addressing those issues with RFID, but only through an infrastructural imagination will people be able to think critically about what various things are saying with their relatively new, networked voices.

Conclusion: The Future of Identification Infrastructures

In the midst of writing this book, I rewatched the science fiction movie *Blade Runner.* The movie was released in 1982 but imagined a future world of 2019. What stood out to me at first was that the makers of *Blade Runner* seemed to imagine a 2019 far more advanced than our present reality. The movie is filled with fully functional androids who have achieved consciousness. In our actual 2018, many of the robots we do have struggle to shuffle across a room. The movie also imagined off-world colonies spread throughout the solar system. In our actual 2018, no human has stepped foot on the moon since 1972.

As I kept watching, however, it dawned on me that in other ways we have progressed far beyond what a movie like *Blade Runner* imagined. We do not have androids or space colonies or flying cars—many places barely have functioning bus systems. However, the world of *Blade Runner* had no internet and no mobile phones. For as influential as *Blade Runner* was in shaping future imaginaries, it was not able to predict two of the technologies that have most affected the way many people live their lives. And nothing about that is unique to *Blade Runner. Back to the Future* had flying cars and hover boards but, once again, no internet. Mark Weiser's predictions about calm computing discussed in chapter 4 were mostly devoid of mobile phones. We might never get our flying cars or personal jet packs, but we did end up with technological advancements few could have predicted.

I mention *Blade Runner* here as a way to touch on the uncertain futures of most technologies. Sometimes predictions are right, but far more often, they are glaringly wrong. As discussed earlier, many smart people writing about the internet predicted people would live in virtual worlds, racism and sexism would lessen as people constructed identity online, and cities would become unimportant because people could live anywhere. What we actually have is an increase in physical mobility, Donald Trump, and a growing global urban population. Many RFID predictions were also wrong, with some experts underestimating the time it would take RFID to become widespread in item-level tagging and overestimating the immediate impact of NFC in mobile phones. To people in the RFID industry in the early 2000s, the present likely looks different than they would have predicted. In addition, to quote novelist William Gibson, "the future is already here—it's just not evenly distributed."[1] For example, item-level tagging has begun to meet some of its predicted retail potential, but it is still limited to only some companies. RFID contactless payment cards are ubiquitous in Europe and Asia but never caught on as much in the United States.

The future is difficult to predict. I wrote my dissertation years ago about a mobile application that later removed the functions I analyzed, so I am no expert in predicting where things are heading. However, in the shorter term, it is possible to make some fairly safe predictions. For instance, barring some major event, the Internet of Things will continue to grow. According to just about every forecast available, the number of objects that connect to the internet or to each other will increase in the coming years. Big data, whether the term survives or not, will continue to expand as more and more companies and institutions quantify everything from purchasing habits to mobility patterns. People will likely write articles about the "failures of big data" (they have already started), and many big data projects will fail.[2] But institutions will continue to find new ways to extract more data out of everything they can, and successful projects will figure out how to use that data productively. Although no one knows exactly what the Internet of Things will look like or what the full consequences of the growth in data collection will be, neither are going away in the near future.

The same predictions can be made about RFID technology more generally. In the next few years, RFID is not going anywhere. The technology in its various forms will continue to be a key piece of identification infrastructure in the Internet of Things. RFID will continue to play a role in transportation systems, track objects through supply chains, and provide data about livestock and pharmaceuticals. The growth of the Internet of Things and data collection will require the differentiation of billions of objects from one another, and RFID will continue to power those processes of sorting and identification. But the next few years will likely see significant shifts in how RFID works as infrastructure to enable object communication and identify billions of things moving through the world. Consequently, this final chapter begins by examining three areas of near-term growth—NFC, item-level tagging, and robot automation—before concluding by returning to RFID's vibrant role as an infrastructure of identification.

RFID and Smartphones as Polymedia

One of the challenges of writing a book about technology is that so much can change so quickly. That kind of change hit in the middle of my research when Apple announced it would partially open up NFC in iOS 11. As discussed in chapter 3, a significant part of what has held NFC back was Apple's decision to restrict NFC to the Apple Pay application. In the United States, phones running iOS constitute between 20 percent and 30 percent of smartphones. In China and the EU5 (Germany, France, Great Britain, Spain, and Italy), iOS is run on about 20 percent of smartphones.[3] Consequently, developers could not rely on people to have NFC access. With Apple's decision to enable more open NFC access, it is quite possible some of the forms of RFID discussed in this book may be replaced by smartphones.

For an obvious example, many of the contactless RFID payment cards popular in Europe, Australia, and Asia may soon give way to mobile payment systems. Mobile payment systems and RFID credit cards are based on many of the same ISO standards, and many payment terminals already accept NFC payments. Transportation infrastructure may also begin to move toward smartphone-based NFC technology. Some cities, such as Barcelona, have already begun letting people pay for public transportation using the NFC in smartphones.[4] In addition, some RFID access systems have moved from RFID badges to NFC-enabled smartphones. Now that NFC is potentially available to almost all smartphone owners, smartphones may become the major way people interact with RFID technologies in the future.

Even with the lifting of Apple's NFC restrictions, a full shift from smart cards and access badges to smartphones will likely be slow. Transportation networks have been running on RFID since the mid 1990s, so even as NFC does become more popular, smartphones will likely exist alongside smart cards for the foreseeable future. After all, not everyone has a smartphone, and smart cards do not have batteries that can die and ruin an evening commute. Consequently, the future of smartphones and NFC will likely work alongside existing RFID infrastructure, just as many types of RFID tagging in retail are used alongside barcodes.

Replacing RFID-enabled cards and badges with smartphones would be significant. It would be a further step toward turning the smartphone into a true "Swiss army knife" device that can be used as everything from a phone to a camera to a credit card to an access badge.[5] But possibly the most significant impact of the potential growth of NFC will be the technology's role in the move toward what Bruce Sterling called the spime.[6] As discussed in chapter 1, spimes are objects that can be tracked temporally and spatially and communicate data to end users. Or, as Julian Bleeker put it in his neologism of the blogject, objects will increasingly be able to tell their own stories to a wider audience than is possible with more proprietary RFID tagging.[7]

We are still years away from objects that fully rise to the criteria Sterling identifies as necessary for the creation of spimes. However, I argue that Apple's opening up of NFC capabilities will likely be an important step toward involving more and more people in the networks of object communication that already exist. Item-level tagging is already happening in the supply chain, but data about objects' mobility is limited mostly to companies. With NFC, smartphones can now become RFID readers for certain tags, and as discussed in chapter 4, some companies have begun using NFC tags to communicate with consumers. It would be surprising if the next few years did not see an increase in NFC tagging designed to inform users about the history of a product. The growth of NFC tagging would then extend many processes of identification to the consumer level, enabling people to identify everything from the vintage on a bottle of wine to the production history of a product. The types of object communication already found in item-level tagging may soon expand as smartphones become an increasingly crucial part of RFID infrastructure.

From 2004 until 2011, NFC mostly floundered. From 2011 until 2017, NFC grew relatively slowly as it became available in most Android phones and as part of Apple Pay in iPhones. Now NFC will be available to most smartphone users, opening up new opportunities for developers to use NFC features and companies to tag items to be read post-purchase. In the near future, many of the forms of identification and object communication covered earlier may focus on smartphones, further ensconcing the devices as polymedia people use to perform the majority of their digital tasks.[8]

Coexisting with Barcodes

RFID's role in item-level tagging has already lived a few full lives, or as one engineer told me, it feels like RFID has already moved through Gartner's stages of the hype cycle multiple times. Item-level tagging was the "next big thing" in the early 2000s, a disappointment in the mid-2000s, and a near-failure by the late 2000s. Then around 2012, RFID tagging began to pick up, and by the mid-2010s industry experts were talking again about a "tipping point" for RFID in retail.[9] The tipping point would be a point where RFID tagging became popular enough that prices would come down further and companies would basically have to adopt RFID or fall behind. When that happens, industry experts predict the rate of RFID adoption in retail will increase exponentially, and the technology could possibly overtake the barcode at some point.[10]

For the foreseeable future, however, barcodes are not going anywhere. Barcodes are a crucial piece of the global economy and one of the most important technologies of the twentieth century. They are established, cheap, and effective producers of identification data. In addition, although barcodes spread quickly in the mid-1970s, they are more an exception than a rule. Many technologies that are ubiquitous today were adopted in fits and starts. For example, mobile phones were commercially available in the 1980s but did not become widely adopted until more than a decade later. The internet remained a fairly niche communication technology throughout the 1980s and early 1990s before becoming the transformative technology it is today. Consequently, while earlier chapters discussed various reasons RFID item-level tagging initially failed to achieve widespread adoption, part of the explanation could be more straightforward. Maybe it was never likely that item-level tagging would move from concept in 1999 to success by the mid-2000s. Now, however, it seems that RFID will play an increasingly important role in retail, though it is doubtful barcodes are going anywhere in the immediate future.

The implications of widespread item-level tagging in retail could be far-reaching. As multiple studies suggest, inventory accuracy will likely increase.[11] A 2018 case study of ten companies that adopted RFID found that adoption led to a 20–25 percent increase in inventory accuracy and a 1.5–5.5 percent increase in sales.[12] All ten companies also reported a return on their RFID investment.[13] As the article notes, although RFID failed to meet its initial hype, multiple case studies and large-scale adoptions suggest that item-level tagging is beginning to establish itself across a variety of industries.

The obvious implication of item-level tagging will be an increase in inventory accuracy. But inventory accuracy is a first step to then enable other practices that could help reshape retail. Below are four likely outcomes as RFID adoption continues across a variety of retail sectors:

- The increased visibility of products on store floors could lead to a wider variety of inventory because stores will need a smaller number of items on the floor. In other words, systems that alert staff when a product is removed could mean fewer of each item would have to be on display, which could lead to a more diverse selection because of the easing of spatial constraints. Something similar happened when widespread use of barcodes led to a "remarkable proliferation of products."[14]
- Barcodes enabled companies to collect increasingly detailed data about purchasing patterns, and RFID would enable even more fine-grained analyses of how products move off shelves.
- The identification data produced through item-level tagging may contribute to new in-store technologies such as magic mirrors or touchscreen displays powered by the identification capabilities of RFID. The case study mentioned above found that multiple companies studied were exploring using the technology in fitting rooms.[15]
- Item-level RFID adoptions will likely help grow omnichannel retailing, which involves creating a "multichannel sales approach that provides the customer with an integrated shopping experience."[16] Omnichannel retailing requires extremely accurate inventory counts, so the growth of RFID adoption will ideally contribute to improved shopping experiences in which people can check online to see exactly what is available in a store.

Finally, the bump in item-level tagging has led to a revival of sorts in talks of the cashier-less store, which I discussed in chapter 6. In the early and mid-2000s, talk began of using RFID to replace cashiers at the grocery store. The idea was that, if every item featured a UHF tag, people could walk by readers while leaving the store and have their purchases logged and their account charged. In 2018, Amazon released its Amazon Go store that removed cashiers, but Amazon insists the store does not use RFID. Others have explored using RFID to build the cashier-less store. The case study of ten large companies that had adopted RFID found that "some are also considering the use of RFID for check-out-less stores,"[17] and a few Chinese companies such as BingoBox have opened smaller-scale stores that use RFID and other technologies to automate the shopping process.

Whether the future of cashier-less stores involves RFID or not, the design will be built on complex processes of identification. For the stores to work, each object must be identifiable and easily readable. The store must be communicative and able to sort through products people remove from shelves. On a deeper level, the processes of identification will contribute to new forms of automation that may have far-reaching consequences. As discussed in chapter 2, unions initially protested barcode adoption because they feared the use of barcodes would lessen the number of cashier jobs. The effects of cashier-less stores—whether through RFID or not—would be far more consequential. According to the Bureau of Labor Statistics, there are around five million retail workers in the United States. If even 20 percent of those jobs were replaced by RFID readers or camera technologies (as with Amazon Go), this change would have a major impact on an already vulnerable segment of the population. The practices of identification mediated through cashiers would be shifted to material infrastructures of identification, and the negative consequences for the vulnerable, who are already often underpaid in retail, could be widespread. Certain jobs, such as some toll collectors, have already been replaced through the automation of RFID. If item-level tagging did ever reach the point at which purchasing could be fully automated, some retail workers could be next.

Of course, for any of the outcomes described above to happen, the public will have to be at least somewhat accepting of—or at least not aware of—RFID tagging. But the reemergence of item-level tagging after some of the failures of the mid-2000s could lead to increased public attention. As chapter 6 examined, RFID tagging plans were met with major concerns by privacy advocates. One can argue that public awareness of the technology was higher in the mid-2000s than it is now, despite the fact that RFID has finally found its way into more retail sites. Some RFID manufacturers do seem to have learned from the privacy uproars of the mid-2000s and have improved privacy measures. One example is the "kill" option that lets retailers permanently disable RFID tags at the point of sale so they cannot be tracked. The problem with the "kill" option is that it can make product returns more complicated because the tag cannot be reactivated. A newer solution is Impinj's Monza R6 tag. The tag is UHF and has a longer read range, but it can be switched to a short read range mode at the point of sale. The tag's read range is then only about 10 percent as far, meaning the tag cannot be read from a distance but can still be read if the object is returned.[18]

Only time will tell what the response will be if RFID item-level tagging becomes more prominent. On the one hand, the industry could see a repeat of some of the boycotts when item-level tagging was explored in the mid-2000s. Almost certainly, groups covered in chapters 4 and 5 will have a reaction to RFID tagging. On the other hand, companies do seem to be more aware of possible issues and may avoid some criticisms by designing more secure systems. Furthermore, the personal data landscape now is more complex than it was in the mid-2000s. To some degree, the data someone could get from identifiable RFID tags may seem a bit quaint compared with the massive data profiles assembled based on mobile phone data, browsing history, credit card purchases, and so on. I am not saying that people should not care about their privacy with RFID tagging. RFID tags can be linked back to personal accounts and tracked to individual people. However, it is possible that the same technology that led to boycotts in the mid-2000s now will be seen as just one small piece of an ever-growing landscape of individual data collection. The public response to RFID tagging—and the larger question of whether there will be a major response at all—will be one of the most interesting questions facing the RFID industry over the next few years.

Robots and Flying Readers

RFID technologies have already replaced human actors in some cases. Fewer people work at tolls because of RFID-enabled toll tags. Fewer people are needed to take inventory at a distribution center if the process is partially automated through RFID. A retail store that uses item-level tagging can take inventory faster and with fewer people. The discussion above looked at the possible impacts if cashier-less stores become a reality. Just as people ceded some agency to barcodes back in the 1970s, they now do so even more with wireless RFID technologies.

Humans might move further into the background of collecting identification data in the coming years. At a growing number of sites, robots and drones are replacing humans as the primary collectors of RFID data. One example can be found in the stores of German clothing chain Adler Modemärkte. The stores use "an RFID-enabled robot called Tory to count inventory and identify the locations of merchandise on store shelves each day."[19] The robot works like a more advanced Roomba, using sensors and navigation software to move through the store floor. Tory is equipped with a UHF reader, so it records tag reads as it moves and further automates the inventory process. According to pilot studies, Tory makes "inventory-counting tasks about 10 times faster than would be possible via a manual count with a handheld reader."[20] European retail giant Tesco has also experimented with using robots to take RFID inventory, and other retailers have also begun using robots similar to the Tory system described above.[21]

Ground-bound robots shuffling through stores are not the only form of automation replacing human actors in RFID systems. Drone technology has also begun to be paired with RFID readers to differentiate objects in large environments. Chapter 4 discussed one such system developed by MIT that uses drones to fly through large warehouse spaces collecting RFID tag reads.[22] Outside warehouses, some research projects have also explored using unmanned robotic vehicles to move through sizable areas powering passive tags and collecting sensor data.[23] As discussed earlier, passively powered wireless sensor networks represent a potential area of growth for RFID technology. They can be deployed over large spaces at comparatively low cost. Once deployed, the reads from the many sensors need to be collected, and that is where the drones come in. The unmanned vehicles in pilot projects move through a space with a UHF reader, powering the sensors and collecting the data. And drones have also been used outside of sensor networks. A Dubai-based company called Age Steel developed a system that uses drones and automates RFID data collection to locate materials in construction sites. One of the managing directors at Age Steel claimed that inventory accuracy in one 950,000-square-foot lot "jumped from around 70 percent to 99.8 percent."[24]

Much of this book has examined at least partially automated forms of identification. UHF RFID can be read quickly from meters away, enabling systems that can read tags as the object moves at regular speeds. The highway toll examples show that many UHF tagged objects can be read without anyone even noticing, which also contributed to some of the privacy concerns surrounding the technology. The insertion of robots and drones into the networks of object communication takes the automation a step further. A tagged object is able to communicate with a drone moving through the environment, showcasing why object communication involves more than direct internet connections. The RFID tag works as a mediator between two objects and turns a warehouse floor or a sensor network into a site of cascading mediators. The tag mediates between object and drone, and the reader on the drone mediates between tag and database. RFID has already automated some processes and played an agential role in networks of object communication. With the pairing of robots and drones and RFID readers, those networks may further remove humans as actors in the process of differentiating objects in the physical world.

Sorting Things Out

The discussions above posited a few likely futures for RFID technology. Maybe the most likely future, however, is that something surprising will happen. As discussed earlier, so much about the successes or failures of RFID as an infrastructure of identification have to do with factors external to the technology. Maybe some country will have a serious disease outbreak in the food chain and then mandate that livestock be tagged with RFID. Maybe the epidemic of counterfeit pharmaceuticals will begin to more seriously affect industrialized nations that receive outsize attention, leading to a rapid growth in RFID pharmaceutical authentication. In those cases, some kind of crisis would emerge that contributed to RFID adoption, and crises almost by definition are difficult to predict.

Regardless of the specific areas of growth or decline, one can be fairly certain that RFID technologies will continue to be a key infrastructure of identification that helps sort the world. After all, people have long sought new ways to differentiate people and things. In their work on the infrastructures of categories and standardization, Geoffrey Bowker and Susan Leigh Star showed how infrastructures of classification work as sorting procedures.[25] These infrastructures, ranging from the complex racial labeling in apartheid South Africa to the International Classification of Diseases, sort phenomena into discrete categories. They differentiate one type from another. This book has built on their work by showing how a ubiquitous, promiscuous mobile technology automates processes of classification and identification. Through tiny tags, materiality becomes sortable in new ways; similarity becomes uniqueness that feeds into larger computing infrastructure. The differentiated "things" might be the rescue mutts sitting patiently—most of the time—at my feet as I type, the car sitting in my driveway, any of the billions of retail items that will be tagged in 2018 alone, or even a human with a microchip implant. On an individual level, the ability to tell one car from another or one pair of jeans from every other pair of jeans may not seem all that revolutionary. However, once the process of sorting and differentiating extends to billions of objects, it becomes something different altogether. The differentiation becomes a key part of the computing embedded into our physical spaces.

The need for granular identification has become a powerful force in the contemporary world. The security state is constantly exploring new ways to differentiate bodies and has worked on systems to uniquely identify people through facial recognition, gait analysis, and other biometric factors. Engineers are improving systems that enable autonomous vehicles to differentiate objects in the physical world to improve automated mobility. RFID fits within the larger trend toward an increasingly fine-grained sorting of objects. Barcodes have long been the king of identification technologies, and for the time being at least, they remain so. But barcodes cannot differentiate billions of individual objects. RFID, on the other hand, can make each object uniquely identifiable, and the examples I have discussed have shown how that differentiation is necessary to animate and power new forms of data production. For billions of things to communicate with the internet and with one another, they have to be differentiated from the voices of all the other things in the environment. RFID technology is, and for the foreseeable future will continue to be, one of the key infrastructures of identification that shape the growing networks of object communication.

RFID tags may not seem as obvious a target for study as mobile technologies like smartphones or tablets. But the embrace of an "infrastructural imagination" encourages looking past more obvious types of communication happening in the physical world. The process may involve peeling back pieces of plastic, reading dry technical standards about data structures, looking at regulations, talking to people in industry, and learning about and experimenting with how RFID technologies work. At some points, looking more deeply into infrastructures can seem overwhelming. For me, as a non-engineer with a job, there is a limit to just how much I can understand about how pieces of RFID work to sort things out. I have read through articles and EPC standards that include sections I do not understand. I have extensively researched pieces of infrastructure only to find out later that they do not communicate through RFID.

Nonetheless, one of the core arguments of this book is that examining contemporary infrastructures of identification is an important part of understanding our world. RFID will, as Nigel Thrift predicted more than a decade ago, form a key part of the technological unconscious that seeks to quantify the world by turning "things" into machine-readable data.[26] The tags will feed into the standardization of space by creating comprehensive logs of where billions of objects are at a given moment. The tags will also work as infrastructures of control through that standardization by opening enclosures and enabling movement in exchange for identification data. As a key part of the technological unconscious, RFID is also part of the vibrant materiality of the physical world. To paraphrase Jane Bennet, the billions of RFID tags moving throughout the world are technological artifacts with "thingly power."[27] The tags play an active role in the environment, whether that role involves letting someone into a building or differentiating one bottle of bourbon from another.

Analyzing practices of identification matters when exploring the various infrastructures that connect objects to the built environment. After all, as media scholar John Durham Peters points out, "Whatever else modernity is, it is a proliferation of infrastructures."[28] RFID technologies are just one relatively small piece of the infrastructures of modernity, but they are an important piece nonetheless. And possibly most importantly, as this book has shown, they are simultaneously a key hardware of the coming Internet of Things and a reminder that the Internet of Things is already here. The physical world is already filled with object communication of various types; people just have to take the time to stop and look.

Notes

1 RFID and the Infrastructural Imagination

1. Robert Hassan, *The Information Society*, Digital Media and Society (Cambridge, UK: Polity, 2008).

2. Here is a link to the GS1 data standard, which includes a section that details the structure of EPC numbers: http://www.gs1.org/epc/tag-data-standard.

3. Klaus Finkenzeller, *RFID Handbook: Fundamentals and Applications in Contactless Smart Cards, Radio Frequency Identification and Near-Field Communication*, trans. Dorte Muller, 1st ed. (London: John Wiley, 1999).

4. Mary Catherine O'Connor, "What You Need to Know about RFID Sensors," *RFID Journal*, January 16, 2012, www.rfidjournal.com/articles/view?9116/4.

5. Lisa Parks and Nicole Starosielski, eds., *Signal Traffic: Critical Studies of Media Infrastructures* (Champaign: University of Illinois Press, 2015); Nicole Starosielski, *The Undersea Network (Sign, Storage, Transmission)* (Durham, NC: Duke University Press, 2015).

6. Susan Leigh Star, "The Ethnography of Infrastructure," *American Behavioral Scientist* 43, no. 3 (1999): 377–391.

7. Jason Farman, "The Materiality of Locative Media," in *Theories of the Mobile Internet: Materialities and Imaginaries*, ed. Andrew Herman, Jan Hadlaw, and Thom Swiss (New York: Routledge, 2015), 45–59.

8. Paul Dourish and Genevieve Bell, "The Infrastructure of Experience and the Experience of Infrastructure: Meaning and Structure in Everyday Encounters with Space," *Environment and Planning B: Planning and Design* 34, no. 3 (2007): 414–430.

9. Nigel Thrift, "Movement-Space: The Changing Domain of Thinking Resulting from the Development of New Kinds of Spatial Awareness," *Economy and Society* 33, no. 4 (2004): 582–604.

10. Brian Larkin, "The Politics and Poetics of Infrastructure," *Annual Review of Anthropology* 42, no. 1 (2013): 327–343.

11. Gerd Kortuem et al., "Smart Objects as Building Blocks for the Internet of Things," *Internet Computing, IEEE* 14, no. 1 (2010): 30–37.

12. Shane Snow, "Foursquare vs. Gowalla: Inside the Check-In Wars," *Fast Company*, March 11, 2010, https://www.fastcompany.com/1578792/foursquare-vs-gowalla-inside-check-wars.

13. Adriana de Souza e Silva and Larissa Hjorth, "Playful Urban Spaces: A Historical Approach to Mobile Games," *Simulation and Gaming* 40, no. 5 (2009): 602–625; Christian Licoppe and Yoriko Inada, "Emergent Uses of a Multiplayer Location-Aware Mobile Game: The Interactional Consequences of Mediated Encounters," *Mobilities* 1, no. 1 (2006): 39–61.

14. Jordan Frith, "Communicating through Location: The Understood Meaning of the Foursquare Check-In," *Journal of Computer-Mediated Communication* 19, no. 4 (2014): 890–905; Lee Humphreys, "Mobile Social Networks and Urban Public Space," *New Media & Society* 12 (2010): 763–778; Christian Licoppe, Carole Anne Rivière, and Julien Morel, "Grindr Casual Hook-Ups as Interactional Achievements," *New Media & Society* 18 (2015): 2540–2558; Michael Saker and Leighton Evans, "Locative Mobile Media and Time: Foursquare and Technological Memory," *First Monday* 21, no. 2 (2016), http://firstmonday.org/ojs/index.php/fm/article/view/6006.

15. Jason Farman, "Map Interfaces and the Production of Locative Media Space," in *Locative Media*, ed. Rowan Wilken and Gerard Goggin (New York: Routledge, 2015), 83–93; Jordan Frith, *Smartphones as Locative Media* (London: Polity, 2015).

16. Carlos Barreneche and Rowan Wilken, "Platform Specificity and the Politics of Location Data Extraction," *European Journal of Cultural Studies* 18, no. 4/5 (2015): 497–513; Mark Graham and Matthew Zook, "Augmented Realities and Uneven Geographies: Exploring the Geolinguistic Contours of the Web," *Environment and Planning A* 45 (2013): 77–99; Rowan Wilken, "Places Nearby: Facebook as a Location-Based Social Media Platform," *New Media & Society* 18, no. 7 (2014): 1087–1103.

17. Adriana de Souza e Silva and Jordan Frith, *Mobile Interfaces in Public Spaces* (New York, NY: Routledge, 2012); Frith, *Smartphones as Locative Media*.

18. Raghu Das, "The Internet of Things and the Potential for RFID," IDTechEx, November 5, 2014, http://www.idtechex.com/research/articles/the-internet-of-things-and-the-potential-for-rfid-00007043.asp; Impinj, "Impinj Sells 10 Billionth Chip as RAIN RFID Goes Mainstream in Retail," Impinj.com, 2015, http://www.impinj.com/about-us/news-and-events/press-releases/2015/impinj-sells-10-billionth-chip-as-rain-rfid-goes-mainstream-in-retail/.

19. IDTechEx, "RFID Forecasts, Players and Opportunities 2017–2027," IDTechEx, 2017, http://www.idtechex.com/research/reports/rfid-forecasts-players-and-opportunities-2017-2027-000546.asp.

20. Geoffrey Bowker and Susan Leigh Star, *Sorting Things Out* (Cambridge, MA: MIT Press, 1999).

21. Stephan Haller, "The Things in the Internet of Things" (Internet of Things Conference, Tokyo, Japan, 2010), http://www.Internet of Things-i.eu/Internet of Things/public/news/resources/TheThingsintheInternetofThings_SH.pdf; Kortuem et al., "Smart Objects as Building Blocks."

22. Pattabhiraman Krishna and David Husak, "RFID Infrastructure," *IEEE Applications & Practice* 45, no. 9 (2007): 4–10; Yuan Lui et al., "An Examination on RFID Innovation Diffusions in Chinese Public Intelligent Transportation Services," *International Journal of Mobile Communications* 13, no. 5 (2015): 549–566.

23. Francisco Borrego-Jaraba et al., "A Ubiquitous NFC Solution for the Development of Tailored Marketing Strategies Based on Discount Vouchers and Loyalty Cards," *Sensors* 13, no. 5 (2013): 6334–6354.

24. Calvin Wong and Z. X. Guo, *Fashion Supply Chain Management Using Radio Frequency Identification (RFID) Technologies* (New York: Woodhead Publishing, 2014); Mikko Keskilammi, Lauri Sydänheimo, and Markku Kivikoski, "Radio Frequency Technology for Automated Manufacturing and Logistics Control. Part 1: Passive RFID Systems and the Effects of Antenna Parameters on Operational Distance," *International Journal of Advanced Manufacturing Technology* 21, no. 10 (2003): 769–774; Siu Keung Kwok and Kenny K. W. Wu, "RFID-Based Intra-supply Chain in Textile Industry," *Industrial Management & Data Systems* 109, no. 9 (2009): 1166–1178.

25. Stephen D. N. Graham, "Software-Sorted Geographies," *Progress in Human Geography* 29, no. 5 (2005): 562–580; David M. Wood and Stephen D. N. Graham, "Permeable Boundaries in the Software-Sorted Society: Surveillance and the Differentiation of Mobility," in *Mobile Technologies of the City*, ed. Mimi Sheller and John Urry (London: Routledge, 2005), 177.

26. Rob Kitchin and Martin Dodge, *Code/Space: Software and Everyday Life* (Cambridge, MA: MIT Press, 2011).

27. Francisco Klauser and Anders Albrechtsund, "From Self-Tracking to Smart Urban Infrastructures: Towards an Interdisciplinary Research Agenda on Big Data," *Surveillance & Society* 12, no. 2 (2014): 273–286.

28. Nigel Thrift, "Remembering the Technological Unconscious by Foregrounding Knowledges of Position," *Environment and Planning D: Society and Space* 22 (2004): 175–190.

29. Nicholas Gane, Couze Venn, and Martin Hand, "Ubiquitous Surveillance: Interview with Katherine Hayles," *Theory, Culture & Society* 24, no. 7/8 (2007): 349–358; N. Katherine Hayles, "RFID: Human Agency and Meaning in Information-Intensive Environments," *Theory, Culture & Society* 26, no. 2/3 (2009): 47–72.

30. Martin Dodge and Rob Kitchin, "Codes of Life: Identification Codes and the Machine-Readable World," *Environment and Planning D: Society and Space* 23, no. 6 (2005): 851–881.

31. Thrift, "Remembering the Technological Unconscious," 185.

32. Hayles, "RFID: Human Agency and Meaning," 48.

33. Hayles, "RFID: Human Agency and Meaning," 66.

34. Dodge and Kitchin, "Codes of Life," 877.

35. Bowker and Star, *Sorting Things Out*, 34.

36. Kenneth Gergen, "The Challenge of Absent Presence," in *Perpetual Contact: Mobile Communication, Private Talk, Public Performance*, ed. James Katz and Mark Aakhus (New York: Cambridge University Press, 2002), 227–241.

37. Amber Case, *Calm Technology: Designing for Billions of Devices and the Internet of Things* (New York: O'Reilly Media, 2016).

38. Kevin Ashton, "That 'Internet of Things' Thing," *RFID Journal*, June 22, 2009, http://www.rfidjournal.com/articles/view?4986.

39. Ian Bogost, *Alien Phenomenology, or What It's Like to Be a Thing* (Minneapolis: University of Minnesota Press, 2012); Graham Harman, *The Quadruple Object* (New York: Zero Books, 2011).

40. Victor Kaptelinin and Bonnie Nardi, *Acting with Technology: Activity Theory and Interaction Design* (Cambridge, MA: MIT Press, 2006).

41. Paul Dourish, *Where the Action Is: The Foundations of Embodied Interaction* (Cambridge, MA: MIT Press, 2001); Maurice Merleau-Ponty, *Phenomenology of Perception*, International Library of Philosophy and Scientific Method (New York: Humanities Press, 1962).

42. Gilles Deleuze and Felix Guattari, *A Thousand Plateaus: Capitalism and Schizophrenia* (Minneapolis: University of Minnesota Press, 1987); Jennifer Daryl Slack and J. Macgregor Wise, *Culture+Technology: A Primer* (New York: Peter Lang, 2005).

43. Jane Bennett, *Vibrant Matter: A Political Ecology of Things* (Durham, NC: Duke University Press, 2010).

44. Madeleine Akrich, "The De-Scription of Technical Objects," in *Shaping Technology/Building Society: Studies in Sociotechnical Change*, ed. Wiebe E. Bijker and John Law (Cambridge, MA: MIT Press, 1992), 205–224; Ignacio Farías and Thomas Bender, eds., *Urban Assemblages: How Actor-Network Theory Changes Urban Studies* (New York: Routledge, 2010); Bruno Latour, *Reassembling the Social: An Introduction to Actor-Network Theory* (Oxford: Oxford University Press, 2005); John Law, "After ANT: Complexity, Naming and Topology," in *Actor Network Theory and After*, ed. John Law and John Hassard (Oxford: Blackwell Publishers, 1999), 1–15.

45. Bruno Latour, *Science in Action: How to Follow Scientists and Engineers through Society* (Cambridge, MA: Harvard University Press, 1987); Bruno Latour, *We Have Never Been Modern* (Cambridge, MA: Harvard University Press, 1993).

46. Bennett, *Vibrant Matter*.

47. Law, "After ANT"; John Law, *After Method: Mess in Social Science Research* (New York: Routledge, 2004).

48. Bennett, *Vibrant Matter*, viii.

49. Latour, *We Have Never Been Modern*.

50. Bennett, *Vibrant Matter*, xiii.

51. RFID tags are not locative media in the same sense as are GPS or cellular triangulation in cell phones. For example, Amal Graafstra argues that "RFID is an identification technology, not a locating technology like GPS, RF beaconing, or mobile phone triangulation." I argue that the difference between a technology of identification and one of location is fairly unimportant. If I move through an electronic toll, the toll identifies that my car was at that location. The same is true for a subway terminal or an access-controlled building. While the tag itself contains only identifying information, once that information is stored with a time stamp by a stationary reader, it essentially becomes location information. See Amal Graafstra, "Presentation 4.3, Dangerousthings.com,"

in *Uberveillance and the Social Implications of Microchip Implantation*, ed. M. G. Michael and Katina Michael (New York: Information Science Reference, 2014), 125–135.

52. Tim Cresswell, "Towards a Politics of Mobility," *Environment and Planning D: Society and Space* 28, no. 1 (2010): 17–31.

53. Mimi Sheller, "Mobility," *Sociopedia.Isa*, 2011, 2, www.sagepub.net/isa/resources/pdf/mobility.pdf.

54. Doreen Massey, *Place, Space, and Gender* (Minneapolis: University of Minnesota Press, 1994).

55. Mimi Sheller and John Urry, "The New Mobilities Paradigm," *Environment and Planning A* 38, no. 2 (2006): 207–226.

56. Bruce Sterling, *Shaping Things* (Cambridge, MA: MIT Press, 2005).

57. Sterling, *Shaping Things*, 11.

58. Julian Bleeker, "A Manifesto for Networked Objects-Cohabiting with Pigeons, Arphids and Aibos in the Internet of Things," Near Future Laboratory, 2006, http://nearfuturelaboratory.com/files/WhyThingsMatter.pdf, n.p.

59. Susan Leigh Star and Karen Ruhleder, "Steps toward an Ecology of Infrastructure: Design and Access for Large Information Spaces," *Information Systems Research* 7, no. 1 (1996): 112.

60. Star and Ruhleder, "Steps toward an Ecology of Infrastructure," 111–134.

61. Larkin, "Politics and Poetics of Infrastructure," 328.

62. Larkin, "Politics and Poetics of Infrastructure," 329.

63. Star, "Ethnography of Infrastructure."

64. Andrew Blum, *Tubes: A Journey to the Center of the Internet* (New York City: Ecco, 2013); Starosielski, *The Undersea Network*.

65. Larkin, "Politics and Poetics of Infrastructure."

66. Blum, *Tubes*; Farman, "Materiality of Locative Media"; Frith, *Smartphones as Locative Media*; Parks and Starosielski, *Signal Traffic*.

67. Stephen D. N. Graham and Simon Marvin, *Splintering Urbanism: Networked Infrastructures, Technological Mobilities, and the Urban Condition* (New York: Routledge, 2001), 11.

68. Brían Hanrahan, "The Anthropoid Condition: Brían Hanrahan Interviews John Durham Peters," *LA Review of Books*, 2015, http://lareviewofbooks.org/article/the-anthropoid-condition-an-interview-with-john-durham-peters/, n.p.; John Durham Peters, *The Marvelous Clouds: Toward a Philosophy of Elemental Media* (Chicago: University of Chicago Press, 2015).

2 Infrastructures of Identification

1. "Fundamental Concepts: Automatic Identification and Data Capture (AIDC) and RFID," GS1, 2012, 3, https://www.gs1us.org/DesktopModules/Bring2mind/DMX/Download.aspx?Command=Core_Download&EntryId=456&language=en-US&PortalId=0&TabId=134.

2. Nigel Thrift, *Knowing Capitalism* (London: Sage, 2005), 220.

3. Nancy Baym, *Personal Connections in the Digital Age* (Cambridge, UK: Polity, 2010), 1.

4. For a more detailed history of the barcode, see Gavin Weightman, "The History of the Bar Code," *Smithsonian*, September 23, 2015, http://www.smithsonianmag.com/innovation/history-bar-code-180956704/.

5. Tony Seideman, "Barcodes Sweep the World," *Wonders of Modern Technology*, Spring 1993, https://www.bar-code.com/upc/bar_code_history.php, n.p.

6. For a detailed history of the development of the UPC, see Stephen Brown, *Revolution at the Checkout Counter: The Explosion of the Bar Code* (Cambridge, MA: Harvard University Press, 1997).

7. Brown, *Revolution at the Checkout Counter*, 42–43.

8. Brown, *Revolution at the Checkout Counter*, 89.

9. The following site explains how the check digit is calculated: http://electronics.howstuffworks.com/gadgets/high-tech-gadgets/upc.htm.

10. Martin Dodge and Rob Kitchin, "Codes of Life: Identification Codes and the Machine-Readable World," *Environment and Planning D: Society and Space* 23, no. 6 (2005): 864.

11. While the thirteen-digit EAN is more popular, there is also an eight-digit EAN standard.

12. Brown, *Revolution at the Checkout Counter*, 199.

13. For a discussion of different types of barcodes, see Francois Lecocq and Cyrille Pepin, "Applications," in *RFID and the Internet of Things*, ed. Herve Chabanne, Pascal Urien, and Jean-Ferdinand Susini (London: Wiley, 2011), 71–112.

14. John T. Dunlop and Jan Rifkin, "Introduction," in *Revolution at the Checkout Counter* (Cambridge, MA: Harvard University Press, 1997), 1–38.

15. Marcus Wohlsen, "Ticker Clocks the Billions of Bar Codes Scanned Each Day," *Wired*, April 12, 2013, https://www.wired.com/2013/04/5-billion-bar-codes-scanned-daily/.

16. Hiromi Hosoya and Markus Schaefer, "Bit Structures," in *Harvard Design School Guide to Shopping*, ed. Chuihua Judy Chung et al. (New York: Taschen, 2002), 157.

17. Adriana de Souza e Silva, "From Cyber to Hybrid: Mobile Technologies as Interfaces of Hybrid Spaces," *Space and Culture* 3 (2006): 261–278; Jordan Frith, *Smartphones as Locative Media* (London: Polity, 2015).

18. Brown, *Revolution at the Checkout Counter*.

19. Jason Gallo and John Laprise, "From Barcodes to RFIDs: Consumer and Commercial Responses to Individual Identification Technologies" (Dresden: International Communication Association, Dresden International Congress Centre, 2013).

20. Dunlop and Rifkin, "Introduction," 12.

21. Chris Potter, "Surveillance Society: Rewards Cards Double as Data Mines," Pittsburgh *Post-Gazette*, June 22, 2015, http://www.post-gazette.com/local/2015/06/22/Rewards-cards-double-as-data-mines/stories/201506220010.

22. James Beniger, *The Control Revolution* (Cambridge, MA: Harvard University Press, 1986).

23. Dunlop and Rifkin, "Introduction," 12–13.

24. Dunlop and Rifkin, "Introduction," 12–13.

25. Alex Kutsishin, "Why QR Codes Don't Work," *Forbes*, August 3, 2012, http://www.forbes.com/sites/ciocentral/2012/08/03/why-qr-codes-dont-work/.

26. Leopoldina Fortunati, "The New Frontiers of Mobile Media: Theoretical Insights on Their Possible Developments," in *Living Inside Mobile Social Information*, ed. James Katz (Boston: Greyden Press, 2014), 60.

27. Fortunati, "New Frontiers of Mobile Media," 62.

28. Shintaro Okazaki, Hairong Li, and Morikazu Hirose, "Benchmarking the Use of QR Code in Mobile Promotion," *Journal of Advertising Research* 52, no. 1 (2012): 102–117.

29. Adam Strout, "The Death of the QR Code," *Marketing Land*, April 4, 2013, http://marketingland.com/the-death-of-the-qr-code-37902.

30. Fortunati, "New Frontiers of Mobile Media," 64.

31. Jacob Beckley, "Heads-Up, Marketers: NFC Will Do More for You than QR Codes," *VentureBeat* (blog), February 17, 2013, https://venturebeat.com/2013/02/17/heads-up-marketers-nfc-will-do-more-for-you-than-qr-codes/.

32. IPv4 was pretty much the only widely used IP standard before IPv6. The first three protocols were experimental versions used in the late 1970s.

33. "RFC: 791. Internet Protocol DARPA Internet Program Protocol Specification" (Internet Engineering Taskforce, 1981), https://tools.ietf.org/html/rfc791.

34. David Goldman, "The Internet Now Has 340 Trillion Trillion Trillion Addresses," CNNMoney, June 6, 2012, http://money.cnn.com/2012/06/06/technology/ipv6/index.htm.

35. Top-level exhaustion of IPv4 addresses occurred in 2011, but individual ISPs had addresses they could still assign and could recycle addresses no longer in use. The lack of addresses also led to the development of new technologies like network address translation that can delay the problem.

36. Confusingly, there is no defined IPv5; that label was skipped to avoid confusion with an experimental protocol called the Internet Stream Protocol, version 2.

37. Ron Hinden, "RFC 1883: Internet Protocol, Version 6 (IPv6) Specification" (Internet Engineering Taskforce, 1995), https://tools.ietf.org/html/rfc1883.

38. Amy Nordrum, "Popular Internet of Things Forecast of 50 Billion Devices by 2020 Is Outdated," *IEEE Spectrum: Technology, Engineering, and Science News,* August 18, 2016, http://spectrum.ieee.org/tech-talk/telecom/internet/popular-internet-of-things-forecast-of-50-billion-devices-by-2020-is-outdated.

39. Iljitsch Van Beijnum, "IPv6 Celebrates Its 20th Birthday by Reaching 10 Percent Deployment," *Ars Technica,* January 3, 2016, https://arstechnica.com/business/2016/01/ipv6-celebrates-its-20th-birthday-by-reaching-10-percent-deployment/.

40. Carolyn Duffy Marsan, "Biggest Mistake for IPv6: It's Not Backwards Compatible, Developers Admit," *Network World,* March 25, 2009, https://www.networkworld.com/article/2265836/lan-wan/biggest-mistake-for-ipv6--it-s-not-backwards-compatible--developers-admit.html.

41. The word "RADAR" is actually an acronym for "radio detection and ranging," coined by the US Navy.

42. For a more detailed discussion of the history of radar, see Massimo Guarnieri, "The Early History of Radar," *IEEE Industrial Electronics Magazine* 4, no. 3 (2010): 36–42.

43. Tony Long, "Feb. 26, 1935: Radar, the Invention That Saved Britain," *Wired,* February 26, 2008, https://www.wired.com/2008/02/dayintech-0226/.

44. As Christoph Rosol points out, Landt's identification of Stockman's paper as the genesis of RFID was picked up in later sources. Those sources repeat Landt's point as if it is uncritical "fact," but as Rosol argues, few articles actually looked into Stockman's article. In addition, as I noted earlier, it is always kind of arbitrary to demarcate a "starting point" in the history of most technologies.

45. Jeremy Landt, "The History of RFID," *Potentials IEEE* 24, no. 4 (2005): 9.

46. Harry Stockman, "Communication by Means of Reflected Power," *Proceedings of the IRE* 36, no. 10 (1948): 1196–1204.

47. Christoph Rosol, "From Radar to Reader: On the Origin of RFID," *Aether* 5 (2010): 40.

48. Stockman, "Communication by Means of Reflected Power," 1196.

49. Landt, "The History of RFID," 9.

50. Landt, "The History of RFID," 9.

51. Stephen Nessen, "Music, Spies and Exact Change: The Strange History of Electronic Tolls," NPR.org, September 15, 2016, http://www.npr.org/sections/alltechconsidered/2016/09/15/487804992/music-spies-and-exact-change-the-strange-history-of-electronic-tolls.

52. Parts of this history are based on Armin Yang, Armin Rida, and Manos Tentzeris, *Design and Development of RFID and RFID-Enabled Sensors on Flexible Low Cost Substrates*, Synthesis Lectures on Computational Electromagnetics S (New York: Morgan and Claypool, 2009), 2–5.

53. Nessen, "Music, Spies and Exact Change."

54. In the United States, early development of RFID technology tended to focus on access control. In the 1980s in Europe, the technology was more widely adopted for animal tracking. See Yang, Rida, and Tentzeris, *Design and Development of RFID*, 3.

55. Ian Wallis, ed., *50 Best Business Ideas That Changed the World* (New York: Jaico Publishing, 2013).

56. "Our History: Memories and Milestones—Exxon and Mobil," Exxon and Mobil, 2017, https://www.exxon.com/en/history.

57. Bob Violino, "The History of RFID Technology," *RFID Journal*, January 16, 2005, http://www.rfidjournal.com/articles/view?1338/2.

58. Finkenzeller, *RFID Handbook: Fundamentals and Applications in Contactless Smart Cards, Radio Frequency Identification and Near-Field Communication* (Chichester, UK: John Wiley, 2010).

59. Kevin Ashton, "That 'Internet of Things' Thing," *RFID Journal*, June 22, 2009, http://www.rfidjournal.com/articles/view?4986.

60. Sanjay Sharma, "A History of the EPC," *SecureIDNews*, July 20, 2005, n.p., https://www.secureidnews.com/news-item/a-history-of-the-epc/.

61. Violino, "History of RFID Technology," 2.

62. Violino, "History of RFID Technology," 2.

63. Sharma, "A History of the EPC."

64. Lawrence Busch, *Standards: Recipes for Reality* (Cambridge, MA: MIT Press, 2011), 3.

65. EPCglobal was created as part of a joint project between the two major barcode standards organizations: GS1 (formerly EAN International) and GS1 US (formerly the Uniform Code Council). EPCglobal is also in charge of other RFID standards, such as the Gen2 UHF air interface, data standards, and reader protocols. EPCglobal now manages the EPCglobal Network that controls EPC numbers and is a spiritual successor to the work done in the early stages of barcode development by the UCC and EAN International.

66. GS1, "EPC Tag Data Standard," GS1, 2017, 14, https://www.gs1.org/sites/default/files/docs/epc/GS1_EPC_TDS_i1_11.pdf.

67. The EPC was not the first data standard developed for RFID. Earlier organizations developed numbering standards for applications such as animal tracking.

68. Companies that use EPC numbers must pay to register with EPCglobal, just like companies had to pay to get a manufacturer code to participate in the UPC system.

69. The numbers in this table come from Dodge and Kitchin, "Codes of Life."

70. Rob Kitchin and Martin Dodge, "Barcodes and RFIDs," in *Globalization in Practice*, ed. Nigel Thrift et al. (Oxford: Oxford University Press, 2014), 271.

71. Adrian Beck, "Measuring the Impact: Key Lessons from 10 Retailers Using RFID" (London: GS1 UK, 2018), 9.

72. "How Do RFID-Based Toll-Collection Systems Work?" *RFID Journal*, September 30, 2013, http://www.rfidjournal.com/blogs/experts/entry?10743, n.p.

73. For a detailed history of the EPC, see Sharma, "A History of the EPC."

74. https://autoidlabs.org/.

75. Sharma, "A History of the EPC," n.p.

76. Bob Violino, "Wal-Mart Expands RFID Mandate," *RFID Journal*, August 17, 2003, http://www.rfidjournal.com/articles/view?539.

77. Matthew Malone, "Did Wal-Mart Love RFID to Death?" *ZDNet*, February 14, 2012, http://www.zdnet.com/article/did-wal-mart-love-rfid-to-death/.

78. "METRO to Penalize RFID Non-Compliance," *RFID Journal*, June 8, 2007, http://www.rfidjournal.com/articles/view?6807; Carol Sliwa, "Target Issues RFID Mandate," *Computer World*, February 27, 2004, http://www.computerworld.com/article/2563853/enterprise-resource-planning/target-issues-rfid-mandate.html; Mary Catherine O'Connor, "DOD Finalizes RFID Mandate Language," *RFID Journal*, September 19, 2005, http://www.rfidjournal.com/articles/view?1876.

79. Bob Violino, "The 5-Cent RFID Tag," *RFID Journal*, January 1, 2004, www.rfidjournal.com/articles/view?748.

80. Malone, "Did Wal-Mart Love RFID to Death?"

81. Malone, "Did Wal-Mart Love RFID to Death?"

82. Malone, "Did Wal-Mart Love RFID to Death?"

83. Malone, "Did Wal-Mart Love RFID to Death?"

84. Mark Roberti, "LIVE! 2015 and the State of RFID," *RFID Journal*, June 24, 2015, https://www.rfidjournal.com/articles/view?13196.

85. Gartner, "Hype Cyle," Gartner Research, 2014, http://www.gartner.com/technology/research/methodologies/hype-cycle.jsp.

86. Beck, "Measuring the Impact," 3.

87. Claire Swedberg, "RFID Brings Lululemon's Inventory Accuracy to 98 Percent," *RFID Journal*, April 18, 2016, http://www.rfidjournal.com/articles/view?14354.

88. Claire Swedberg, "American Apparel Deploys Real-Time, Storewide RFID Inventory-Management Solution," *RFID Journal*, August 13, 2013, http://www.rfidjournal.com/articles/view?10906.

89. "Marks & Spencer Will Tag 400M Items in 2014," RFID24-7.com, February 27, 2014, http://rfid24-7.com/2014/02/27/marks-will-tag-400m-items-in-2014/.

90. Mark Roberti, "Divining the True State of RFID Adoption in Retail," *RFID Journal*, December 16, 2015, https://www.rfidjournal.com/articles/view?13862.

91. Barbara Thau, "Is the 'RFID Retail Revolution' Finally Here? A Macy's Case Study," *Forbes*, May 15, 2017, https://www.forbes.com/sites/barbarathau/2017/05/15/is-the-rfid-retail-revolution-finally-here-a-macys-case-study/#632b0d933294.

3 Understanding RFID Technologies

1. Jordan Frith, *Smartphones as Locative Media* (London: Polity, 2015).

2. From a business standpoint, the way RFID systems are assembled and sold is very different from that of smartphones. Smartphones consist of many pieces of hardware, but the actual phone is put together and sold by one manufacturer. To put together a full RFID system often requires buying tags from one company, readers from another company, and software from another company. The lack of a dominant player who

sells complete RFID solutions was a repeated source of frustration with industry people I interviewed.

3. Manish Bhuptani and Sharam Moradpour, *RFID Field Guide: Deploying Radio Frequency Identification Systems* (New York: Prentice Hall, 2006).

4. Ken Traub, "Big RFID Data Software," *RFID Journal*, July 19, 2015, http://www.rfidjournal.com/articles/view?13286.

5. Sarah Smith, "RFID Forecasts, Players and Opportunities 2016–2026," IDTechEx, November 15, 2016, http://www.prnewswire.com/news-releases/rfid-forecasts-players-and-opportunities-2016-2026-300179309.html.

6. Bob Violino, "The Basics of RFID Technology," *RFID Journal*, January 16, 2005, http://www.rfidjournal.com/articles/view?1337/3, n.p.

7. "How Much Do RFID Tags Cost Today?" *RFID Journal*, FAQ 2016, https://www.rfidjournal.com/faq/show?85.

8. LF RFID typically operates at 125 kHz and occasionally at 134 kHz.

9. James Heurich, "The Myths of LF vs. HF," *RFID Journal*, December 21, 2009, http://www.rfidjournal.com/articles/view?7278.

10. Mark Roberti, "Do All Countries Use the Same Frequencies?" *RFID Journal*, FAQ 2016, https://www.rfidjournal.com/faq/show?62.

11. Roberti, "Do All Countries Use the Same Frequencies?" n.p.

12. Dan Newling, "Britons 'Could Be Microchipped like Dogs in a Decade,'" *Daily Mail*, October 30, 2006, http://www.dailymail.co.uk/news/article-413345/Britons-microchipped-like-dogs-decade.html.

13. Jane Bennett, *Vibrant Matter: A Political Ecology of Things* (Durham, NC: Duke University Press, 2010).

14. Vivian Tan, "What Is the Difference between MIFARE Card and RFID Card?" LinkedIn Pulse, 2016, https://www.linkedin.com/pulse/what-difference-between-mifare-card-rfid-vivian.

15. "About Smart Cards: Frequently Asked Questions," Smart Card Alliance, 2017, http://www.smartcardalliance.org/smart-cards-faq/, n.p.

16. MIFARE technology is one example. MIFARE is trademarked by NXP Semiconductors and refers to a few variations of chips used in contact and contactless payments. Some sources differentiate between RFID and MIFARE, but in reality, MIFARE's contactless cards use the same ISO standards as RFID contactless payment cards. In other words, MIFARE is a type of RFID, just a type that includes more complexity than a read-only EPC tag.

17. Karl Koscher et al., "EPC RFID Tags in Security Applications: Passport Cards, Enhanced Drivers Licenses, and Beyond," 2008, ftp://ftp.cs.washington.edu/tr/2008/10/UW-CSE-08-10-02.PDF.

18. Someone could be tracked through the personal serial number on their card, but that could already happen in much easier ways through smartphone data, CCTV, toll tag data, and so on.

19. ID Stronghold, "How Crooks Are Stealing Credit Card Information," YouTube video, 2011, https://www.youtube.com/watch?v=x3S_6EJCjn0.

20. Will Oremus, "The Skimming Scam," *Slate*, August 25, 2015, http://www.slate.com/articles/life/travel_explainer/2015/08/credit_cards_passports_and_rfid_fraud_are_special_blocking_wallets_necessary.html.

21. "FACT CHECK: Electronic Pickpocketing," Snopes.com, 2012, https://www.snopes.com/fact-check/electronic-pickpocketing/, n.p.

22. Sumit Roy et al., "RFID: From Supply Chains to Sensor Nets," *Proceedings of the IEEE* 98, no. 9 (2010): 1583–1592.

23. Jun Zhang et al., "A Review of Passive RFID Tag Antenna-Based Sensors and Systems for Structural Health Monitoring Applications," *Sensors* 17, no. 2 (2017): 1–33.

24. Claire Swedberg, "Passive Sensor Tags to Surpass 5 Million Units This Year," *RFID Journal*, November 14, 2017, www.rfidjournal.com/articles/view?16809.

25. Priyanka Rawat et al., "Wireless Sensor Networks: A Survey on Recent Developments and Potential Synergies," *Journal of Supercomputing* 68, no. 1 (2014): 1–48, esp. 2.

26. Rawat et al., "Wireless Sensor Networks," 2.

27. Zhang et al., "A Review of Passive RFID Tag," 265.

28. Zhang et al., "A Review of Passive RFID Tag."

29. Rob Dunn, "RFID versus NFC: The Big Difference You Should Know About," LinkedIn Pulse, 2015, https://www.linkedin.com/pulse/rfid-versus-nfc-big-difference-you-should-know-rob-dunn.

30. Lauren Taylor, "What's the Difference between NFC and RFID?" *Event Farm* (blog), November 2, 2016, http://blog.eventfarm.com/blog/whats-the-difference-between-nfc-and-rfid.

31. James Thrasher, "RFID versus NFC: What's the Difference?" *RFID Insider* (blog), October 11, 2013, http://blog.atlasrfidstore.com/rfid-vs-nfc.

32. ECMA, "ECMA Welcomes ISO/IEC Adoption of NFC Standard for Short Range Wireless Communication," ECMA International, press release, March 18, 2004, http://ecma-international.org/flat/news/Ecma-340-NFCIP-1.htm, n.p.

33. ECMA, "ECMA Welcomes ISO/IEC Adoption," n.p.

34. Judith Vanderkay, "Nokia, Philips and Sony Establish the Near Field Communication (NFC) Forum," NFC Forum, 2004, https://nfc-forum.org/newsroom/nokia-philips-and-sony-establish-the-near-field-communication-nfc-forum/.

35. Jonathan Collins, "Could NFC Fail to Take Off?" *RFID Journal*, February 15, 2008, http://www.rfidjournal.com/articles/view?3916, n.p.

36. Sarah Clark, "GSMA Calls for Mass Market NFC Handsets by Mid-2009," NFC World, November 19, 2008, https://www.nfcworld.com/2008/11/19/3235/gsma-calls-for-mass-market-nfc-handsets-by-mid-2009/.

37. Google has had kind of a tortured history of naming its mobile payment application. The original name was Google Wallet. Then it changed the name of the pay app to Android Pay, and Google Wallet became an app to send money to peers. Then, in 2018, it changed Android Pay to Google Pay.

38. Rian Boden, "More than 100m People Will Make an NFC Mobile Payment in 2016," NFC World, March 19, 2016, https://www.nfcworld.com/2016/03/10/343180/more-than-100m-people-will-make-an-nfc-mobile-payment-in-2016/.

39. Samsung Pay is slightly more complicated because it uses both NFC and Magnetic Secure Transmission (MST). The MST is used to basically trick terminals that accept only magnetic stripe cards into accepting Samsung Pay.

40. Karthik Raghavan, "Apple Pay vs Google Wallet: How They Work," Investopedia, January 2, 2015, https://www.investopedia.com/articles/personal-finance/010215/apple-pay-vs-google-wallet-how-they-work.asp.

41. "Use Any Credit or Debit Card with Google Wallet," *Google Commerce*, August 1, 2012, https://commerce.googleblog.com/2012/08/use-any-credit-or-debit-card-with.html, n.p.

42. Megan Geuss, "Android Pay Is All About Tokenization; Google Wallet Takes a Backseat," *Ars Technica*, May 28, 2015, https://arstechnica.com/information-technology/2015/05/android-pay-will-embrace-tokenization-mostly-replace-google-wallet/.

43. Kif Leswing, "7 Years after Steve Jobs Waged War on Flash, It's Officially Dying," *Business Insider*, July 25, 2017, http://www.businessinsider.com/adobe-flash-killed-by-2020-2017-7.

44. AppleInsider, "Apple to Limit iPhone 6 NFC to Apple Pay, Restrict Developer Access," AppleInsider, September 16, 2014, //appleinsider.com/articles/14/09/16/apple-to-limit-iphone-6-nfc-to-apple-pay-restricts-developer-access.

45. Benjamin Mayo, "Apple Says Opening Up iPhone NFC Would 'Fundamentally Diminish' Security as Australian Banks Resist Apple Pay," 9to5Mac, August 10, 2016, https://9to5mac.com/2016/08/10/apple-says-opening-up-iphone-nfc-would-fundamentally-diminish-security-as-australian-banks-resist-apple-pay/.

46. Claire Swedberg, "Australian Banks Challenge Apple's Closed NFC," *RFID Journal*, February 21, 2017, http://www.rfidjournal.com/articles/view?15703.

47. John Foley Jr., "How to Utilize NFC for Print Marketing," *Printing Impressions*, February 26, 2015, http://www.piworld.com/post/utilize-nfc-print-marketing/.

48. "DYNE X NFC," Dyne Menswear, 2017, https://dyne.life/pages/dyne-x-nfc.

49. Evan Schuman, "Apparel Maker Meshes RFID, NFC and QR Together—and Makes It All Work," *ComputerWorld*, April 8, 2016, http://www.computerworld.com/article/3053547/retail-it/apparel-maker-meshes-rfid-nfc-and-qr-together-and-makes-it-all-work.html.

50. Rian Boden, "French Vineyard to Safeguard Its Entire Output with NFC," NFC World, June 21, 2016, https://www.nfcworld.com/2016/06/21/345687/french-vineyard-safeguard-entire-output-nfc/.

51. Thinfilm Printed Electronics, "Diageo and Thinfilm Unveil the Connected Smart Bottle Printed Electronics Technology," YouTube video, 2015, https://www.youtube.com/watch?v=7myP2_KLumc.

52. Andrew Dalton, "iOS 11 Could Use the iPhone's NFC Chip for More than Apple Pay," Engadget, June 6, 2017, https://www.engadget.com/2017/06/06/ios-11-iphone-core-nfc-chip-more-than-apple-pay/.

53. Rian Boden, "NXP Issues Statement of Support for NFC Tag Reading on Apple Devices," NFC World, June 9, 2017, https://www.nfcworld.com/2017/06/09/353194/nxp-issues-statement-of-support-for-nfc-tag-reading-on-apple-devices/.

54. Bruno Latour, *Pandora's Hope: Essays on the Reality of Science Studies* (Cambridge, MA: Harvard University Press, 1999), 303.

55. Ian Bogost, *Alien Phenomenology, or What It's Like to Be a Thing* (Minneapolis: University of Minnesota Press, 2012).

4 RFID and the Internet of Things

1. "No More There," YouTube video, 1997, http://www.youtube.com/watch?v=ioVMoeCbrig&feature=youtube_gdata_player.

2. Paul Virilio, *Open Sky* (London: Verso, 1997).

3. Kevin Morgan, "The Exaggerated Death of Geography: Learning, Proximity, and Territorial Systems," *Journal of Economic Geography* 4 (2004): 3–21.

4. Nicholas Negroponte, *Being Digital* (New York: Vintage Books, 1995).

5. The idea of "cyberspaces" in popular fiction still exists. The 2011 novel *Ready Player One*, for example, was a best seller and focused on the idea of bodies fleeing into virtual reality.

6. Andrew Blum, *Tubes: A Journey to the Center of the Internet* (New York City: Ecco, 2013).

7. Nancy Baym, *Personal Connections in the Digital Age*, 2nd ed. (Cambridge, UK: Polity, 2015).

8. N. Katherine Hayles, *How We Became Posthuman: Virtual Bodies in Cybernetics, Literature, and Informatics* (Chicago: University of Chicago Press, 1999).

9. Samuel Greengard, *The Internet of Things* (Cambridge, MA: MIT Press, 2015), 15.

10. Stroud, "Internet of Things (IoT) Definition—Webopedia," Webopedia, 2017, https://www.webopedia.com/TERM/I/internet_of_things.html.

11. Luigi Aztori, Antonio Iera, and Giacomo Morabito, "The Internet of Things: A Survey," *Computer Networks* 54, no. 15 (2010): 2787–2805.

12. Steven Weber and Richmond Y. Wong, "The New World of Data: Four Provocations on the Internet of Things," *First Monday* 22, no. 2 (2017), http://firstmonday.org/ojs/index.php/fm/article/view/6936, n.p.

13. Eduardo Aibar and Wiebe E. Bijker, "Constructing a City: The Cerdà Plan for the Extension of Barcelona," *Science, Technology & Human Values* 22, no. 1 (1997): 3–30.

14. H. Lefebvre, *The Production of Space* (Malden, MA: Blackwell, 1991).

15. Yi Fu Tuan, *Space and Place: The Perspective of Experience* (Minneapolis: University of Minnesota Press, 1977), 136.

16. Nigel Thrift, *Non-Representational Theory: Space, Politics, Affect* (London: Routledge, 2007), 91.

17. Mike Crang and Stephen Graham, "Sentient Cities: Ambient Intelligence and the Politics of Urban Space," *Information, Communication & Society* 10, no. 6 (2007): 789–817; Rob Kitchin and Martin Dodge, *Code/Space: Software and Everyday Life* (Cambridge, MA: MIT Press, 2011).

18. Nigel Thrift, "Movement-Space: The Changing Domain of Thinking Resulting from the Development of New Kinds of Spatial Awareness," *Economy and Society* 33, no. 4 (2004): 582–604.

19. Rob Kitchin and Martin Dodge, "Airport/Code Spaces," in *Aeromobilities*, ed. Saulo Cwerner, Sven Kesserling, and John Urry (London: Routledge, 2009), 96.

20. Kitchin and Dodge, *Code/Space*.

21. "RFID Guns," Smart Tech Challenges Foundation, 2016, https://smarttechfoundation.org/smart-firearms-technology/rfid/.

22. Lyle Bateman, "Explaining RFID Technology and Poker," *Poker News*, January 30, 2014, https://ca.pokernews.com/news/2014/01/explaining-rfid-technology-and-poker-2999.htm.

23. Robert Vamosi, "Gone in 60 Seconds—the High-Tech Version," CNET, 2006, https://www.cnet.com/news/gone-in-60-seconds-the-high-tech-version/.

24. The National Rifle Association (NRA) is opposed to smart gun technology in part because of fears of creeping government control and in part because they are the NRA. See Robert J. Szcerba, "The NRA's Next Battleground…Smart Guns," *Forbes*, January 11, 2016. https://www.forbes.com/sites/robertszczerba/2016/01/11/the-nras-next-battleground-smart-guns/#747dec8b4d7f.

25. Samuel Greengard, *The Internet of Things* (Cambridge, MA: MIT Press, 2015).

26. Nigel Thrift, "Remembering the Technological Unconscious by Foregrounding Knowledges of Position," *Environment and Planning D: Society and Space* 22 (2004): 175–190.

27. Nigel Thrift, "Remembering the Technological Unconscious," 183.

28. Nigel Thrift, "Space: The Fundamental Stuff of Geography," in *Key Concepts in Geography*, ed. Nicholas Clifford et al., 2nd ed. (London: Sage, 2009), 87.

29. Dave Michels, "'IoT' Is Just a Fad," *TalkingPointz*, September 13, 2016, https://talkingpointz.com/Internet of Things-is-just-a-fad/.

30. Amy Nordrum, "Popular Internet of Things Forecast of 50 Billion Devices by 2020 Is Outdated," *IEEE Spectrum: Technology, Engineering, and Science News*, August 18, 2016, http://spectrum.ieee.org/tech-talk/telecom/internet/popular-internet-of-things-forecast-of-50-billion-devices-by-2020-is-outdated.

31. Ole Jensen, "Flows of Meaning, Cultures of Movements: Urban Mobility as Meaningful Everyday Life Practice," *Mobilities* 4, no. 1 (2009): 139–158.

32. Mimi Sheller and John Urry, "The New Mobilities Paradigm," *Environment and Planning A* 38, no. 2 (2006): 207–226.

33. Crang and Graham, "Sentient Cities."

34. Stephen Graham and Simon Marvin, *Splintering Urbanism: Networked Infrastructures, Technological Mobilities, and the Urban Condition* (New York: Routledge, 2001).

35. David Wood and Stephen Graham, "Permeable Boundaries in the Software-Sorted Society: Surveillance and the Differentiation of Mobility," in *Mobile Technologies of the City*, ed. Mimi Sheller and John Urry (London: Routledge, 2005), 183.

36. Jordan Frith, "Splintered Space: Hybrid Spaces and Differential Mobility," *Mobilities* 7, no. 1 (2012): 131–149.

37. David Lyon, "Surveillance, Security and Social Sorting: Emerging Research Priorities," *International Criminal Justice Review* 17, no. 3 (2007): 161–170.

38. Rian Boden, "Barcelona Introduces Contactless and NFC Mobile Payments across Public Transport Network," NFC World, November 13, 2015, https://www.nfcworld.com/2015/11/13/339571/barcelona-introduces-contactless-and-nfc-mobile-payments-across-public-transport-network/;

Adrian Lim, "Commuters Can Use Compatible NFC Phones to Pay Bus and Train Fares," *Straits Times*, March 29, 2016, http://www.straitstimes.com/singapore/transport/commuters-can-use-compatible-nfc-phones-to-pay-bus-and-train-fares.

39. Chandana Unnithan, "Examining Innovation Translation of RFID Technology in Australian Hospitals through a Lens Informed by Actor-Network Theory," PhD diss., Victoria University, Victoria, BC, 2014.

40. Ari Juels, David Molnar, and David Wagner, "Security and Privacy Issues in E-Passports," in *Proceedings of the First International Conference on Security and Privacy for Emerging Areas in Communications Networks (SECU-RECOMM'05)* (Washington, DC: IEEE Computer Society, 2005), 74–88.

41. Ayse Ceyhan, "Technologization of Security: Management of Uncertainty and Risk in the Age of Biometrics," *Surveillance & Society* 5, no. 2 (2008): 102–123.

42. Kevin Hannam, Mimi Sheller, and John Urry, "Mobilities, Immobilities and Moorings," *Mobilities* 1, no. 1 (2006): 1–22.

43. James Beniger, *The Control Revolution* (Cambridge, MA: Harvard University Press, 1986).

44. Jeremy Landt, "The History of RFID," *Potentials IEEE* 24, no. 4 (2005): 9; Christoph Rosol, "From Radar to Reader: On the Origin of RFID," *Aether* 5 (2010): 40.

45. Claire Swedberg, "Boeing Tags Shipments to the DOD," *RFID Journal*, May 19, 2005, http://www.rfidjournal.com/articles/view?1587.

46. "Marks & Spencer Expands RFID Rollout; Will Tag 400M Items in 2014," RFID 24-7, February 27, 2014; Swedberg, "American Apparel Deploys Real-Time, Storewide RFID Inventory-Management Solution," *RFID Journal*, August 13, 2013.

47. Bill Hardgrave, "Tracking Your Competitors," *RFID Journal*, October 15, 2012, https://www.rfidjournal.com/articles/view?10020.

48. This Lululemon video shows how quickly inventory can be taken with RFID: https://www.youtube.com/watch?v=cZfx2naKYXo.

49. Hardgrave, "Tracking Your Competitors"; Swedberg, "RFID Brings Lululemon's Inventory Accuracy to 98 Percent," *RFID Journal*, April 18, 2016.

50. Dani Deahl, "MIT Figured Out a Better Way for Drones to Use RFID Technology," *The Verge*, September 2, 2017, https://www.theverge.com/2017/9/2/16217302/mit-drones-billion-dollar-problem-inventory-rfid-tags.

51. Laxmiprasad Putta, "RF Enabled Shelf and Back Room," Auto-ID Center, 2002, http://cryptome.org/rfid/savant.pdf.

52. Joan Melià-Seguí et al., "Enhancing the Shopping Experience through RFID in an Actual Retail Store," in *Proceedings of the 2013 ACM Conference on Pervasive and Ubiquitous Computing Adjunct Publication*, UbiComp '13 Adjunct (New York: ACM, 2013), 1029–1036.

53. Here is a video from a Ralph Lauren store showing how a magic mirror can work: https://www.youtube.com/watch?v=vFF95SvTfRE.

54. Unnithan, "Examining Innovation Translation of RFID Technology."

55. Abraham Blonder, "Why Zone-Based Real-Time Location Systems Are Superior," *RFID Journal*, August 22, 2011, http://www.rfidjournal.com/articles/view?8583.

56. Claire Swedberg, "ORLocate RFID-Enabled System for Surgical Sponges and Instruments Gets FDA Clearance," *RFID Journal*, August 27, 2010, www.rfidjournal.com/articles/view?7836.

57. Genevieve Bell and Paul Dourish, "Yesterday's Tomorrows: Notes on Ubiquitous Computing's Dominant Vision," *Personal and Ubiquitous Computing Journal* 11, no. 2 (2006): 133–143.

58. Ruth Schwartz Cowan, *More Work for Mother: The Ironies of Household Technology from the Open Hearth to the Microwave* (New York: Basic Books, 1985).

59. Francis K. Aldrich, "Smart Homes: Past, Present and Future," ed. Richard Harper (New York: Springer, 2003), 17–40.

60. Aldrich, "Smart Homes," 17.

61. James Titcomb, "Internet of Things Struggles as Use of Smart Home Gadgets Flatlines," *The Telegraph*, August 27, 2016, http://www.telegraph.co.uk/technology/2016/08/27/internet-of-things-struggles-as-use-of-smart-home-gadgets-flatli/.

62. Mohsen Darianian and Martin Peter Michael, "Smart Home Mobile RFID-Based Internet-of-Things Systems and Services," in *2008 International Conference on Advanced Computer Theory and Engineering* (Phuket, Thailand, 2008), 116–120; Ayeesha Hafeez et al., "Smart Home Area Networks Protocols within the Smart Grid Context," *Journal of Communications* 9, no. 9 (2014): 665–671.

63. Gerd Kortuem et al., "Smart Objects as Building Blocks for the Internet of Things," *IEEE Internet Computing* 14, no. 1 (2010): 44–51.

64. Sumi Helal et al., "The Gator Tech Smart House: A Programmable Pervasive Space," *Computer* 38, no. 3 (2005): 50–60; Tatsuya Yamazaki, "Beyond the Smart Home," in *2006 International Conference on Hybrid Information Technology* 2 (2006): 350–355.

65. To clear up terminology, at some point, someone put a touchscreen on a refrigerator and called it a smart fridge. Those smart fridges are different from RFID-enabled smart fridges.

66. "Connected Home Collaboration Surface," Corning, 2016, https://www.corning.com/worldwide/en/innovation/the-glass-age/the-glass-age-today/see-you-in-the-kitchen.html.

67. "RFID/NFC Enabled Smart Washing Machine Detects Fabric, Supports Remote Maintenance," NXP, 2012, https://nxp-rfid.com/rfidnfc-enabled-smart-washing-machine-detects-fabric-supports-remote-maintenance/.

68. Helal et al., "The Gator Tech Smart House."

69. Tomás Sánchez López et al., "Taxonomy, Technology and Applications of Smart Objects," *Information Systems Frontiers* 13, no. 2 (2011): 281–300.

70. Aldrich, "Smart Homes," 18.

71. Anne-Jorunn Berg, Cynthia Chew, and Ruza Furst-Dilic, "A Gendered Socio-Technical Construction: The Smart House," in *Bringing*

Technology Home: Gender and Technology in a Changing Europe (Buckingham: Open University Press, 1994), 160.

72. Cowan, *More Work for Mother.*

73. "Microchipping of Animals," American Veterinary Medical Association, 2013, https://www.avma.org/KB/Resources/Reference/Pages/Microchipping-of-Animals-Backgrounder.aspx.

74. Biohacking tends to refer to amateur practices of body modification, and "the term body hacking, also known as biohacking, emerged in the 90s alongside the dotcom revolution, and is generally used by a cyberpunk wing of the body modification scene." For more details, see Lissette Olivares, "Hacking the Body and Posthumanist Transbecoming: 10,000 Generations Later as the Mestizaje of Speculative Cyborg Feminism and Significant Otherness," *NanoEthics* 8, no. 3 (2014): 288.

75. Donna Haraway, "A Cyborg Manifesto: Science, Technology, and Socialist-Feminism in the Late Twentieth Century," in *Simians, Cyborgs and Women: The Reinvention of Nature* (New York: Routledge, 1991), 149.

76. Steve Connor, "Professor Has World's First Silicon Chip Implant," *The Independent,* August 26, 1998, http://www.independent.co.uk/news/professor-has-worlds-first-silicon-chip-implant-1174101.html.

77. Bárbara Nascimento Duarte, "Entangled Agencies: New Individual Practices of Human-Technology Hybridism through Body Hacking," *NanoEthics* 8, no. 3 (2014): 275–285; Nancy Nisbet, "Resisting Surveillance: Identity and Implantable Microchips," *Leonardo* 37, no. 3 (2004): 210–214.

78. Will Weissert, "Microchips Implanted in Mexican Officials," msnbc.com, July 14, 2004, http://www.nbcnews.com/id/5439055/ns/technology_and_science-tech_and_gadgets/t/microchips-implanted-mexican-officials/.

79. "Barcelona Clubbers Get Chipped," BBC News, September 29, 2004, http://news.bbc.co.uk/2/hi/technology/3697940.stm.

80. Dan Newling, "Britons 'Could Be Microchipped like Dogs in a Decade,'" *Daily Mail,* October 30, 2006.

81. Amal Graafstra, "Amal Graafstra: Technologist, Author and Double RFID Implantee," Dangerous Things, 2017, http://amal.net/, n.p.

82. The videos can be found on the Dangerous Things YouTube channel: https://www.youtube.com/channel/UCiMeqSKIPTvyOMZe1n4Pp4A.

83. "Dangerous Things," Dangerous Things, 2017, https://dangerousthings.com/.

84. Bonnie Burton, "Swedish Office Gets Under Employees' Skin with RFID Microchips," CNET, February 3, 2015, http://www.cnet.com/news/swedish-office-gets-under-employees-skin-with-rfid-microchips/.

85. Jim Edwards, "The Controversy Magnet: PositiveID 'Chips' Alzheimer's Patients, Quite Possibly without Permission," CBS News, March 18, 2010, http://www.cbsnews.com/news/the-controversy-magnet-positiveid-chips-alzheimers-patients-quite-possibly-without-permission/.

86. Borut Werber and Anja Žnidaršič, "The Use of Subcutaneous RFID Microchip in Health Care: A Willingness to Challenge," *Health and Technology* 5, no. 1 (2015): 57–65.

87. Chris Weller, "A Swedish Rail Line Now Scans Microchip Implants in Addition to Accepting Paper Tickets," *Business Insider*, June 20, 2017, https://www.businessinsider.in/A-Swedish-rail-line-now-scans-microchip-implants-in-addition-to-accepting-paper-tickets/articleshow/59226279.cms.

88. Merrit Kennedy, "Wisconsin Company Offers to Implant Chips in Its Employees," NPR.org, July 25, 2017, http://www.npr.org/sections/thetwo-way/2017/07/25/539265157/wisconsin-company-plans-to-start-implanting-chips-in-its-employees, n.p.

89. In my research, I have come across peer-reviewed academic articles that unironically use Prison Planet as a source when discussing RFID. Prison Planet is not a good source.

90. Mac Slavo, "Hotel California: L.A. Officials to Send Homeless to Internment Camps: Will Be Implanted with RFID Chips," InfoWars, July 29, 2014, https://www.infowars.com/hotel-california-l-a-officials

-to-send-homeless-to-internment-camps-will-be-implanted-with-rfid-chips/.

91. Darrin McBreen, "666: It Has Begun!" InfoWars, March 6, 2015, https://www.infowars.com/666-it-has-begun/.

92. The Alex Jones Channel, "Swedish Gov. Announces Plan to Microchip Entire Population / Genders to Be Declared Illegal," YouTube video, 2017, https://www.youtube.com/watch?v=E-emCBC_Axw.

93. Steve Watson, "RFID Tags in New US Notes Explode When You Try to Microwave Them," Prison Planet, 2005, https://www.prisonplanet.com/022904rfidtagsexplode.html.

94. Paul Joseph Watson, "RFID Microchips to Be Embedded in Breast Implants," InfoWars, 2013, https://www.infowars.com/rfid-microchips-to-be-embedded-in-breast-implants/.

95. "FALSE: Australia Becomes First Country to Begin Microchipping Its Public," Snopes.com, 2016, https://www.snopes.com/australia-becomes-first-country-to-begin-microchipping/.

96. "RFID Chip Implemented in All Public Schools by 2015: Effort to Curb Gun Violence," Snopes.com, 2014, https://www.snopes.com/media/notnews/rfidgun.asp.

97. "Study Finds 1 in 3 Americans Have Been Implanted with RFID Chips: Most Unaware," Snopes.com, 2014, https://www.snopes.com/politics/satire/microchip2.asp.

98. Imbeffie, *NBC Prediction That We Will All Have an RFID Chip Under Our Skin by 2017*, YouTube video, 2007, https://www.youtube.com/watch?v=1YJsxMcAJoA&feature=youtu.be.

99. David Vose, "NBC ANNOUNCES EVERYONE MUST WEAR A MICROCHIP BY 2017," YouTube video, 2016, https://www.youtube.com/watch?v=volHNGg2UUY.

100. David Mikkleson, "FALSE: U.S. Residents to Be Implanted with Microchips," Snopes.com, 2010, http://www.snopes.com/politics/medical/microchip.asp.

101. Mark Roberti, "Proof Obamacare Requires All Americans to Be Chipped," *RFID Journal*, October 22, 2013, http://www.rfidjournal.com/blogs/rfid-journal/entry?10775.

102. Jefferson Graham, "You Will Get Chipped—Eventually," *USA Today*, August 9, 2017, https://www.usatoday.com/story/tech/2017/08/09/you-get-chipped-eventually/547336001/, n.p.

103. If you are one of those people, see Landru613, "Subcutaneous Microchip Removal," YouTube video, 2013, https://www.youtube.com/watch?v=0yKNGVCcMkw.

104. Weber and Wong, "The New World of Data," n.p.

105. N. Katherine Hayles, "RFID: Human Agency and Meaning in Information-Intensive Environments," *Theory, Culture & Society* 26, no. 2/3 (2009): 47–72.

106. Mark Weiser and John Seely Brown, "Designing Calm Technology," Ubiq.com, 1995, http://www.ubiq.com/hypertext/weiser/acmfuture2endnote.htm.

107. Mark Weiser, "The Computer for the 21st Century," *Scientific American* 265, no. 3 (1991): 95.

108. Weiser and Brown, "Designing Calm Technology," n.p.

109. Amber Case, *Calm Technology: Designing for Billions of Devices and the Internet of Things* (New York: O'Reilly Media, 2016).

110. Paul Dourish and Genevieve Bell, *Divining a Digital Future: Mess and Mythology in Ubiquitous Computing* (Cambridge, MA: MIT Press, 2014).

111. Jordan Frith, *Smartphones as Locative Media* (London: Polity, 2015).

112. Mark Weiser, "The World Is Not a Desktop," *ACM Interactions* 1, no. 1 (1994): 7.

113. Case, *Calm Technology*.

114. Matthew Chalmers, "A Historical View of Context," *Computer Supported Cooperative* 13 (2004): 223–247.

115. Josh Mchugh, "Attention, Shoppers: You Can Now Speed Straight Through Checkout Lines!" *Wired,* July 1, 2004, https://www.wired.com/2004/07/shoppers/.

5 Data Traces of Identification

1. Robert Malone, "Tracking Sushi," *Forbes,* May 23, 2007, https://www.forbes.com/2007/05/23/logistics-restaurants-biz-logistics-cx_rm_0523sushi.html.

2. Michael Belfiore, "Solving NASA's Water Problem," *RFID Journal,* July 13, 2015, https://www.rfidjournal.com/articles/view?13260.

3. Ray Y. Zhong et al., "Visualization of RFID-Enabled Shopfloor Logistics Big Data in Cloud Manufacturing," *International Journal of Advanced Manufacturing Technology* 84, no. 1 (2016): 5–16, esp. 6.

4. Viktor Mayer-Schönberger and Kenneth Cukier, *Big Data: A Revolution That Will Transform How We Live, Work, and Think* (New York: Eamon Dolan/Houghton Mifflin Harcourt, 2013), 8.

5. Chris Anderson, "The End of Theory: The Data Deluge Makes the Scientific Method Obsolete," *Wired,* June 23, 2008, http://archive.wired.com/science/discoveries/magazine/16-07/pb_theory.

6. Paul Zikopoulos et al., *Understanding Big Data* (New York: McGraw Hill, 2012).

7. David McCandless, "Data, Information, Knowledge, Wisdom?" Information is Beautiful, November 29, 2010, http://www.informationisbeautiful.net/2010/data-information-knowledge-wisdom/.

8. Rob Kitchin, *The Data Revolution: Big Data, Open Data, Data Infrastructures and Their Consequences* (London: Sage, 2014), 21.

9. Lisa Gitelman, ed., *Raw Data Is an Oxymoron* (Cambridge, MA: MIT Press, 2013).

10. Bruno Latour, *Science in Action: How to Follow Scientists and Engineers through Society* (Cambridge, MA: Harvard University Press, 1987).

11. Kitchin, *The Data Revolution*, 21.

12. Mayer-Schönberger and Cukier, *Big Data*.

13. Michael E. Hobart and Zachary S. Schiffman, *Information Ages: Literacy, Numeracy, and the Computer Revolution* (Baltimore, MD: Johns Hopkins University Press, 1998).

14. James Beniger, *The Control Revolution* (Cambridge, MA: Harvard University Press, 1986).

15. Zikopoulos et al., *Understanding Big Data*.

16. The three Vs are not the only way of conceptualizing big data. In his book *The Data Revolution*, Rob Kitchin defines seven qualities of big data to expand on the more commonly used three Vs. To quote his work, the seven criteria are that data must be

- huge in *volume*, consisting of terabytes or petabytes of data;
- high in *velocity*, being created in or near real-time;
- diverse in *variety*, being structured and unstructured in nature;
- *exhaustive* in scope, striving to capture entire populations or systems (n=all);
- fine-grained in *resolution* and uniquely *indexical* in identification;
- *relational* in nature, containing common fields that enable the conjoining of different data sets;
- *flexible*, holding the traits of *extensionality* (can add new fields easily) and *scaleability* (can expand in size rapidly).

Quoted from Kitchin, *The Data Revolution*, 68.

17. Anderson, "The End of Theory"; Mayer-Schönberger and Cukier, *Big Data*.

18. Mayer-Schönberger and Cukier, *Big Data*, 13.

19. Mayer-Schönberger and Cukier, *Big Data*, 7.

20. Anderson, "The End of Theory," n.p.

21. Gitelman, *Raw Data Is an Oxymoron*; Kitchin, *The Data Revolution*.

22. danah boyd and Kate Crawford, "Critical Questions for Big Data," *Information, Communication & Society* 15, no. 5 (2012): 662–679.

23. Kitchin, *The Data Revolution*, 7.

24. For an extensive discussion of the MagicBand, see Cliff Kuang, "Disney's $1 Billion Bet on a Magical Wristband," *Wired*, March 10, 2015, http://www.wired.com/2015/03/disney-magicb and/.

25. Kevin Seifert, "NFL Teams Prepping for RFID Data Dump," ESPN, 2016, http://www.espn.com/blog/nflnation/post/_/id/200005.

26. Claire Swedberg, "RFID Helps Soccer Teams Keep Their Eye on the Ball, and Their Players," *RFID Journal*, March 13, 2012, http://www .rfidjournal.com/articles/view?9315.

27. Mark Roberti, "RFID Helps Analyze Hockey Players' Performance," *RFID Journal*, August 26, 2014, https://www.rfidjournal.com/articles /view?12118.

28. Rayoman Engineer, "Unique RFID Tags to Help Reunite Lost Children with Their Families at Sabarimala," The Better India, December 13, 2017, https://www.thebetterindia.com/124181/vodafone-kerala -police-rfid-tags-sabarimala/.

29. Roozbeh Derakhshan, Maria E. Orlowska, and Xue Li, "RFID Data Management: Challenges and Opportunities," in *2007 IEEE International Conference on RFID*, January 1, 2007, 175–182, https://trove.nla.gov.au /work/3812400?q&versionId=49472305.

30. Federico Guerrini, "Cities Cannot Be Reduced to Just Big Data and IoT: Smart City Lessons From Yinchuan, China," *Forbes*, September 19, 2016, https://www.forbes.com/sites/federicoguerrini/2016/09/19/engaging -citizens-or-just-managing-them-smart-city-lessons-from-china /#21786eddab0f.

31. Mariko Hirose, "Newly Obtained Records Reveal Extensive Monitoring of E-ZPass Tags throughout New York," American Civil Liberties Union,

April 24, 2015, https://www.aclu.org/blog/free-future/newly-obtained-records-reveal-extensive-monitoring-e-zpass-tags-throughout-new-york.

32. "Flowing City," The Flowing City Project, 2016, http://flowingcity.com/ n.p.

33. Michel Foucault, *Discipline and Punish* (New York: Pantheon Books, 1977).

34. Gilles Deleuze, "Postscript on the Society of Control," *October* 59 (Winter) (1992): 7.

35. Bill Hardgrave, "Omnichannel Retailing," *RFID Journal*, December 18, 2012, https://www.rfidjournal.com/articles/view?10247.

36. Kitchin, *The Data Revolution.*

37. Manish Bhuptani and Sharam Moradpour, *RFID Field Guide: Deploying Radio Frequency Identification Systems* (New York: Prentice Hall, 2006).

38. Ken Traub, "Big RFID Data Software," *RFID Journal*, July 19, 2015, http://www.rfidjournal.com/articles/view?13286.

39. Susan Flake, "RFID Earns a Second Chance," *RFID Journal*, October 1, 2011, https://www.rfidjournal.com/articles/view?8824; Ken Traub, "EPCIS for Internal Projects," *RFID Journal*, November 4, 2013, https://www.rfidjournal.com/articles/view?11148.

40. Anderson, "The End of Theory," n.p.; Mayer-Schönberger and Cukier, *Big Data*, 14.

41. Lawrence Busch, *Standards: Recipes for Reality* (Cambridge, MA: MIT Press, 2011).

42. Mark Roberti, "Divining the True State of RFID Adoption in Retail," *RFID Journal*, December 16, 2015.

43. Matthew Malone, "Did Wal-Mart Love RFID to Death?" *ZDNet*, February 14, 2012.

44. Ashton Morrow, "Delta Introduces Innovative Baggage Tracking Process," *Delta News Hub*, April 28, 2016, http://news.delta.com/delta-introduces-innovative-baggage-tracking-process-0.

45. Claire Swedberg, "Men's Clothing Store Brings RFID Intelligence to Fitting Room," *RFID Journal*, 2018, http://www.rfidjournal.com/articles/view?17135/2.

46. iPourIt, "FAQs," iPourIt, Inc. (blog), 2018, http://www.ipouritinc.com/faq/.

47. IDTechEx, "RFID Forecasts, Players and Opportunities 2017–2027," https://www.idtechex.com/research/reports/rfid-forecasts-players-and-opportunities-2017-2027-000546.asp.

48. "Bioterrorism Act of 2002: Title III—Protecting Safety and Security of Food and Drug Supply" (Washington, DC: US Federal Drug Administration, 2002), SEC 307.

49. Theodoros Varzakas, "ISO 22000, HACCP, and Other Management Tools for Implementation of Food Safety-Traceability Case Studies," in *Handbook of Food Processing*, ed. Theodoros Varzakas and Constantina Tiza (Boca Raton, FL: CRC Press, 2016), 105–140.

50. Jeremy Landt, "The History of RFID," *Potentials IEEE* 24, no. 4 (2005): 9.

51. "FAQ," Canadian Cattle Identification Agency (CCIA), 2017, https://www.canadaid.ca/about_us/faqs.html.

52. "RFID Reader Standards, Procedures, and Testing Document Radio Frequency Identification (RFID). Version 8" (Alberta: Canadian Cattle Identification Agency, 2013), 3, https://canadaid.ca/about_us/documents/CCIA_Reader_Standards_Version_8-April%202013_Final_Eng.pdf.

53. "Tagging Livestock," Meat and Livestock Australia, 2016, https://www.mla.com.au/meat-safety-and-traceability/red-meat-integrity-system/about-the-national-livestock-identification-system-2015/tagging-livestock/.

54. "Mandatory Cattle Identification Program Q & A," Department of Agriculture and Rural Development, 2015, http://www.michigan.gov/mdard/0,4610,7-125-48096_48149-137059—,00.html.

55. "Official Animal Identification Number (AIN) Devices" (Washington, DC: US Department of Agriculture, 2009).

56. Sandrine Berthaud-Clair, "Radio-Tagging Sheep Will Spell the Death of the Old Ways, Say French Farmers," *The Guardian*, September 22, 2014, https://www.theguardian.com/environment/2014/sep/22/french-farmers-sheep-radio-tagging.

57. "Swine Identification 101: Understanding Swine Ear Tag Requirements," Texas Pork, 2014, https://texaspork.org/content/uploads/2014/10/Swine_ID.pdf.

58. Virginia Harrison, "Holy Cow! India Is the World's Largest Beef Exporter," CNNMoney, 2015, http://money.cnn.com/2015/08/05/news/economy/india-beef-exports-buffalo/index.html.

59. Rose Troup Buchanan, "Mad Cow Disease in the UK: What Is BSE and What Are the Symptoms?" *The Independent*, October 1, 2015, http://www.independent.co.uk/news/uk/home-news/mad-cow-disease-in-the-uk-what-is-bse-and-what-are-the-symptoms-a6675351.html.

60. Kristy E. Boehler, "The Evolution of Animal Identification in Beef Cattle," *Nebraska Law Review* 91, no. 3 (2013): 567–599.

61. Boehler, "The Evolution of Animal Identification," 571–572.

62. Boehler, "The Evolution of Animal Identification," 581.

63. "Review of Selected Cattle Identification and Tracing Systems Worldwide," MAF Biosecurity New Zealand Information Paper No: 2009/03 (Auckland, NZ: MAF Biosecurity New Zealand, 2009), iii.

64. Berthaud-Clair, "Radio-Tagging Sheep."

65. "Global Cold Chain RFID Market Size, Share, Development, Growth and Demand Forecast to 2022," P&S Market Research, 2016, https://www.psmarketresearch.com/market-analysis/rfid-in-cold-chain-market.

66. The cold chain also pertains to pharmaceuticals. Possibly the most developed example of RFID in the pharmaceutical cold chain is a recent rollout by the Japanese Suzuken Group. The Suzuken Group deployed one thousand RFID-enabled coolers at various sites, including hospitals and pharmacies. The coolers are equipped with readers that detect the tags on medicines when they are placed inside. The readers also capture temperature data transmitted from the tags on medicines,

enabling a comprehensive monitoring system of cold chain data. See Claire Swedberg, "Suzuken Group Deploys RFID-Enabled Coolers for Drug Tracking," *RFID Journal*, February 7, 2017, http://www.rfidjournal.com/articles/view?15588.

67. Estefania Abad et al., "RFID Smart Tag for Traceability and Cold Chain Monitoring of Foods: Demonstration in an Intercontinental Fresh Fish Logistic Chain," *Journal of Food Engineering* 93, no. 4 (2009): 394–399.

68. Kwanho Kim et al., "I-RM: An Intelligent Risk Management Framework for Context-Aware Ubiquitous Cold Chain Logistics," *Expert Systems with Applications* 46 (2016): 463–473.

69. Steven Weber and Richmond Wong, "The New World of Data: Four Provocations on the Internet of Things," *First Monday* 22, no. 2 (2017).

70. Alexandra Ossola, "The Fake Drug Industry Is Exploding, and We Can't Do Anything About It," *Newsweek*, September 25, 2015, http://www.newsweek.com/2015/09/25/fake-drug-industry-exploding-and-we-cant-do-anything-about-it-373088.html.

71. "2016 Top Markets Report Pharmaceuticals" (Washington, DC: International Trade Administration, 2017), http://trade.gov/topmarkets/pdf/Pharmaceuticals_Executive_Summary.pdf.

72. Ossola, "Fake Drug Industry Is Exploding."

73. Paul Toscano, "The Dangerous World of Counterfeit Prescription Drugs," CNBC, 2011, http://www.cnbc.com/id/44759526.

74. "More Counterfeit Lipitor Found in UK," *Pharmaceutical Journal*, August 19, 2005, http://www.pharmaceutical-journal.com/news-and-analysis/more-counterfeit-lipitor-found-in-uk/10020443.article.

75. Jocelyne Sambira, "Counterfeit Drugs Raise Africa's Temperature," *Africa Renewal*, May 2013, http://www.un.org/africarenewal/magazine/may-2013/counterfeit-drugs-raise-africa%E2%80%99s-temperature.

76. "Survey of the Quality of Selected Antimalarial Medicines Circulating in Six Countries of Sub-Saharan Africa" (Geneva: World Health Organization, 2011), http://www.who.int/medicines/publications/WHO_QAMSA_report.pdf?ua=1.

77. For a brief history, see Douglas Taylor, "RFID in the Pharmaceutical Industry: Addressing Counterfeits with Technology," *Journal of Medical Systems* 38 (2014): 140–145.

78. Taylor, "RFID in the Pharmaceutical Industry."

79. JD Howard, "Implementation of RFID in the Pharmaceutical Industry" (San Luis Obispo: California Polytechnic, 2009), https://www.iopp.org/files/public/CAPolytechJDHoward.pdf.

80. Matthieu P. Schapranow et al., "Costs of Authentic Pharmaceuticals: Research on Qualitative and Quantitative Aspects of Enabling Anti-Counterfeiting in RFID-Aided Supply Chains," *Personal and Ubiquitous Computing* 16, no. 3 (2012): 271–289.

81. John Ayoade, "RFID Adoption in the Developed and Developing World," *Journal of Current Issues in Media & Telecommunications* 3, no. 1 (2011): 73–83.

82. Alberto Coustasse et al., "Could the Pharmaceutical Industry Benefit from Full-Scale Adoption of Radio-Frequency Identification (RFID) Technology with New Regulations?" *Perspectives in Health Information Management* 13 (Fall 2016): 1b.

83. Wietse Van Ransbeek, "Why the Blockchain Could Transform the Face of Digital Democracy," *CitizenLab*, 2017, https://www.citizenlab.co/blog/civic-tech/blockchain-could-transform-democracy/.

84. Roger Aitken, "Blockchain to the Rescue Creating a 'New Future' for Digital Identities," *Forbes*, January 7, 2018, https://www.forbes.com/sites/rogeraitken/2018/01/07/blockchain-to-the-rescue-creating-a-new-future-for-digital-identities/.

85. "Bananacoin—Join Organic Token Generation Event," Bananacoin, 2018, https://bananacoin.io/.

86. Evelyn Cheng, "'Long Island Iced Tea' Micro-Cap Adds Blockchain to Name and Stock Soars," CNBC, December 21, 2017, https://www.cnbc.com/2017/12/21/long-island-iced-tea-micro-cap-adds-blockchain-to-name-and-stock-soars.html.

87. Fitz Tepper, "SEC Warns against Public Companies Adding Blockchain to Their Name," *TechCrunch*, January 25, 2018, http://social.techcrunch.com/2018/01/25/sec-warns-against-public-companies-adding-blockchain-to-their-name/.

88. Steve Banker, "Blockchain in the Supply Chain: Too Much Hype," *Forbes*, September 1, 2017, https://www.forbes.com/sites/stevebanker/2017/09/01/blockchain-in-the-supply-chain-too-much-hype/.

89. Deloitte, "Continuous Interconnected Supply Chain: Using Blockchain and Internet-of-Things in Supply Chain Traceability" (Deloitte, 2017), https://www2.deloitte.com/content/dam/Deloitte/lu/Documents/technology/lu-blockchain-internet-things-supply-chain-traceability.pdf.

90. TransparentPlanet, "Blockchain+RFID=Total Product Lifecycle Management" (RAIN RFID Alliance Meeting, Seattle, WA, 2017), https://rainrfid.org/wp-content/uploads/2017/05/5-TransparentPlanet3-FINAL.pdf.

91. Feng Tian, "An Agri-Food Supply Chain Traceability System for China Based on RFID and Blockchain Technology," in *2016 13th International Conference on Service Systems and Service Management (ICSSSM)*, 2016, 1–6.

92. "Can Blockchain Technology Put an End to Counterfeit Drugs?" *Drug Patent Watch* (blog), 2017, https://www.drugpatentwatch.com/blog/can-blockchain-technology-put-an-end-to-counterfeit-drugs/.

93. Possibly the most unique thing about Waltonchain is that the currency lost a huge chunk of its value because of a silly social media mistake: https://mashable.com/2018/02/28/waltonchain-twitter-scam-wtc/#2UVuk1My3qq6.

94. Rob Marvin, "23 Weird, Gimmicky, Straight-Up Silly Cryptocurrencies," *PCMAG*, February 6, 2018, https://www.pcmag.com/feature/358046/23-weird-gimmicky-straight-up-silly-cryptocurrencies.

95. Revelation 13:16–17.

96. Roger Barrier, "The Mark of the Beast and RFID Chips Analysis—Revelations," Preach It, Teach It, 2016, https://www.preachitteachit.org/ask-roger/detail/rfid-chips-the-mark-of-the-beast/.

97. Katherine Albrecht and Liz Mcintyre, *The Spychips Threat: Why Christians Should Resist RFID and Electronic Surveillance* (New York: Thomas Nelson, 2006).

98. Kim Peckham, "Satan's Spy Chips," HopeChannel, 2016, https://www.hopechannel.com/read/satan-s-spy-chips; Maratha Trumpeter, "Vital Facts about the Mark of the Beast (RFID Chip/Human Barcode: 666)," Rapture Watcher, 2012, https://rapturewatcher.wordpress.com/vital-facts-about-the-mark-of-the-beast-rfid-chip-human-barcode-666/.

99. Marshall Connoly, "Would You Accept the MARK OF THE BEAST to Get a Job?" Catholic Online, 2017, https://www.catholic.org/news/technology/story.php?id=74365.

100. Coustasse et al., "Could the Pharmaceutical Industry Benefit?"

101. Helen Zhao, "Bitcoin Blockchain Consumes a Lot of Energy. Engineers are Changing That," CNBC, February 23, 2018, https://www.cnbc.com/2018/02/23/bitcoin-blockchain-consumes-a-lot-of-energy-engineers-changing-that.html.

102. N. Katherine Hayles, "RFID: Human Agency and Meaning in Information-Intensive Environments," *Theory, Culture & Society* 26, no. 2/3 (2009): 47–72.

6 Surveillance and the Mobility of Bodies

1. Mark Roberti, "Addressing Fears about RFID," *RFID Journal*, October 1, 2008, www.rfidjournal.com/articles/view?4395.

2. European Commission, "Privacy and Data Protection Impact Assessment Framework for RFID Applications" (Brussels: European Commission, 2011); Federal Trade Commission, "Radio Frequency Identification: Applications and Implications for Consumers" (Washington, DC: Federal Trade Commission, 2005).

3. Katherine Albrecht and Liz McIntyre, *Spychips: How Major Corporations and Government Plan to Track Your Every Purchase and Watch Your Every Move* (New York: Plume, 2005).

4. "Smart Card Technology FAQ," Smart Card Alliance, 2016, http://www.smartcardalliance.org/smart-cards-faq/.

5. Anders Albrechtslund, "Online Social Networking as Participatory Surveillance," *First Monday* 13, no. 3 (2008); d. boyd, "Facebook's Privacy Trainwreck," *Convergence: The International Journal of Research into New Media Technologies* 14, no. 1 (2008): 13–20; B. Debatin et al., "Facebook and Online Privacy: Attitudes, Behaviors, and Unintended Consequences," *Journal of Computer-Mediated Communication* 15 (2009): 83–108.

6. Alistair Beresford and Frank Stajano, "Location Privacy in Pervasive Computing," *IEEE Pervasive Computing* 2, no. 1 (2003): 46–55; Andrew J. Blumberg and Peter Eckersley, "On Locational Privacy and How to Avoid Losing It Forever," Electronic Frontier Foundation, 2009, https://www.eff.org/wp/locational-privacy; Adriana de Souza e Silva and Jordan Frith, "Locational Privacy in Public Spaces: Media Discourses on the Personalization and Control of Space by Location-Aware Mobile Media," *Communication, Culture, & Critique* 3, no. 4 (2010): 503–525.

7. Joseph Turow, "Americans and Online Privacy: The System Is Broken" (University of Pennsylvania: Annenberg Public Policy Center, 2003), http://www.securitymanagement.com/archive/library/Anneberg_privacy1003.pdf; Joseph Turow, *The Daily You: How the New Advertising Industry Is Defining Your Identity and Your Worth* (New Haven, CT: Yale University Press, 2012).

8. Sean P. Hier, "Probing the Surveillant Assemblage: On the Dialectics of Surveillance Practices as Processes of Social Control," *Surveillance and Society* 1, no. 3 (2003): 399–411; Francisco Klauser and Anders Albrechtslund, "From Self-Tracking to Smart Urban Infrastructures: Towards an Interdisciplinary Research Agenda on Big Data," *Surveillance and Society* 12, no. 2 (2014); David Lyon, *The Electronic Eye: The Rise of Surveillance Society* (Minneapolis: University of Minnesota Press, 1994).

9. Kevin D. Haggerty and Richard V. Ericson, "The Surveillant Assemblage," *British Journal of Sociology* 51 (2000): 605–622.

10. David Lyon, *Surveillance after Snowden* (London: Polity, 2015), 1.

11. Quoted from European Commission, "DG Connect Internal Report on the Implementation of the Commission Recommendation on the Implementation of Privacy and Data Protection Principles in Applications Supported by Radio-Frequency Identification" (Brussels: European Commission, 2014), 2.

12. David Lyon, "Liquid Surveillance: The Contribution of Zygmunt Bauman to Surveillance Studies," *International Political Sociology* 4 (2010): 325–338.

13. Zygmunt Bauman, *Liquid Modernity* (London: Polity, 2000).

14. danah boyd, "Debating Privacy in a Networked World for the WSJ," danah boyd, Apophenia, November 10, 2011, http://www.zephoria.org/thoughts/archives/2011/11/20/debating-privacy-in-a-networked-world-for-the-wsj.html.

15. Kate Raynes-Goldie, "Aliases, Creeping, and Wall Cleaning: Understanding Privacy in the Age of Facebook," *First Monday* 15, no. 1 (2010), http://firstmonday.org/ojs/index.php/fm/article/view/2775/2432.

16. Daniel Castro and Alan McQuinn, "The Privacy Panic Cycle: A Guide to Public Fears about New Technologies" (Information Technology and Innovation Foundation, 2015), www2.itif.org/2015-privacy-panic.pdf.

17. The Spychips website was taken down at some point in 2017. It now directs people to a new website: CAMCAT—Citizens Against Marking, Chipping and Tracking, www.camcat.org.

18. Bob Violino, "Behind the Benetton Brouhaha," *RFID Journal*, April 13, 2003, http://www.rfidjournal.com/articles/view?381.

19. Alorie Gilbert, "Wal-Mart Cancels 'Smart Shelf' Trial," CNET, 2003, http://www.cnet.com/news/wal-mart-cancels-smart-shelf-trial/.

20. BBC, "Tesco 'Spychips' Anger Consumers," BBC, 2005, http://news.bbc.co.uk/2/hi/business/4209545.stm; Jack Neff, "Privacy Group Slams Levi's for RFID-Chip Clothing Tags," *Ad Age*, April 28, 2006, http://adage.com/article/news/privacy-group-slams-levi-s-rfid-chip-clothing-tags/108846/.

21. The open letter can be found at https://www.aclu.org/sites/default/files/FilesPDFs/icaoletter.pdf.

22. The National Conference of State Legislatures has a good collection of RFID-related laws in the United States: http://www.ncsl.org/research/telecommunications-and-information-technology/radio-frequency-identification-rfid-privacy-laws.aspx.

23. Dan Newling, "Britons 'Could Be Microchipped Like Dogs in a Decade,'" *Daily Mail*, October 30, 2006, www.dailymail.co.uk/news/article-413345/Britons-microchipped-like-dogs-decade.html.

24. James Katz and Ronald Rice, "Is RFID Feared? Consumer Sentiments towards RFID-Like Healthcare Communication Technology" (International Communication Association, Montreal, Canada, 2008).

25. Katz and Rice, "Is RFID Feared?" 3.

26. Albrecht and McIntyre, *Spychips*, 60.

27. Katherine Albrecht, "Katherine Albrecht: Bio," Spychips.com, 2006, http://www.spychips.com/katherine-albrecht.html.

28. Mark Roberti, "Spychips Book Fails to Make Its Case," *RFID Journal*, October 24, 2005, http://www.rfidjournal.com/articles/view?1947.

29. A representative quote: "We know that a Big Brother vision of the future sounds farfetched. We didn't believe it ourselves until we saw with our own eyes and heard with our own ears companies detailing their mind-boggling plans" (Albrecht and McIntyre, *Spychips*, 3).

30. Albrecht and McIntyre, *Spychips*, 215, 169.

31. Albrecht and McIntyre, *Spychips*, 211.

32. Castro and McQuinn, "The Privacy Panic Cycle."

33. Nancy Baym, *Personal Connections in the Digital Age* (Cambridge, UK: Polity, 2015); Carolyn Marvin, *When Old Technologies Were New: Thinking about Electric Communication in the Late Nineteenth Century* (Oxford: Oxford University Press, 1988).

34. Jenifer Sunrise Winter, "Surveillance in Ubiquitous Network Societies: Normative Conflicts Related to the Consumer In-Store Supermarket Experience in the Context of the Internet of Things," *Ethics and Information Technology* 16, no. 1 (2013): 27–41.

35. Jeremy Gillula and Dave Maass, "California: Protect Your Driver License Privacy," Electronic Frontier Foundation, 2015, https://www.eff.org/deeplinks/2015/09/california-protect-your-drivers-license-privacy.

36. European Commission, "Privacy and Data Protection Impact Assessment."

37. Jay Stanley, "Newest School RFID Scheme Is Reminder of Technology's Surveillance Potential," American Civil Liberties Union, 2012, https://www.aclu.org/blog/newest-school-rfid-scheme-reminder-technologys-surveillance-potential.

38. "Information Stored by Oyster System—a Freedom of Information Request to Transport for London," WhatDoTheyKnow, 2013, https://www.whatdotheyknow.com/request/information_stored_by_oyster_sys.

39. "Cards Let Metro Collect Data on Riders, Track Trips," *Washington Times*, May 17, 2005, www.washingtontimes.com/news/2005/may/17/20050517-120301-3752r/.

40. Lyon, "Liquid Surveillance," 325.

41. Blumberg and Eckersley, "On Locational Privacy"; K. W. Ogden, "Privacy Issues in Electronic Toll Collection," *Implications of New Information Technology* 9, no. 2 (2001): 123–134.

42. Yunus Kathawala and Benjamin Tueck, "The Use of RFID for Traffic Management," *International Journal of Technology, Policy and Management* 8, no. 2 (2008).

43. Hirose, "Newly Obtained Records Reveal Extensive Monitoring of E-ZPass Tags throughout New York," American Civil Liberties Union, April 24, 2015, https://www.aclu.org/blog/privacy-technology/location-tracking/newly-obtained-records-reveal-extensive-monitoring-e-zpass.

44. Hirose, "Newly Obtained Records Reveal," n.p.

45. Hirose, "Newly Obtained Records Reveal," n.p.

46. Karen Louise Smith et al., "Playing with Surveillance: The Design of a Mock RFID-Based Identification Infrastructure for Public Engagement," *Surveillance & Society* 9, no. 1/2 (2011): 149–166.

47. Ellen Balka and Susan Leigh Star, "Mapping the Body across Diverse Information Systems: Shadow Bodies and How They Make Us Human," in *Boundary Objects and Beyond: Working with Susan Leigh Star*, ed. Geoffrey C. Bowker et al. (Cambridge, MA: MIT Press, 2015), 424.

48. Kevin Ashton, "Coming Together for a Common Goal," *RFID Journal*, April 15, 2012, www.rfidjournal.com/articles/view?9423.

49. Gilbert, "Wal-Mart Cancels 'Smart Shelf' Trial."

50. John Hind, M. Mathewson James, and Marcia Peters, Identification and Tracking of Persons Using RFID in Store Environments, patent filed 2001 and issued 2006, 1, https://www.google.com/patents/US7076441.

51. Hind, James, and Peters, Identification and Tracking, 8.

52. Hind, James, and Peters, Identification and Tracking, 8.

53. Hind, James, and Peters, Identification and Tracking, 7.

54. Jeremy Otto and Dennis Seitz, Automated Monitoring of Activity of Shoppers in a Market, US 20020113123 A1, patent filed 2000, and issued 2002.

55. Turow, *The Daily You.*

56. RFID hygiene monitoring technology is not new. The technology was first explored in the mid-2000s by companies such as Woodward Laboratories that developed a technology called the iHygiene Perfect Pump.

57. Amy Lipton, "RFID Helps Hospitals Clean Up Their Act," *RFID Journal*, September 28, 2014, www.rfidjournal.com/articles/view?12211.

58. Employee location monitoring is not a new phenomenon and not exclusive to RFID. One of the earliest and most famous examples was the Active Badge system. The system did not use RFID, but it was an

important precursor to some of the RFID systems in existence today, and it raised privacy concerns from employees. Discussing the system, computer scientist Paul Dourish explained how he refused to use the system while he was an employee at Xerox PARC because he did not want his movements tracked. Many people who cannot access areas or material without RFID badges do not have that choice. See Paul Dourish and Genevieve Bell, *Divining a Digital Future* (Cambridge, MA: MIT Press, 2011).

59. Stan Kurkovsky, "Continuous RFID-Enabled Authentication: Privacy Implications," *IEEE Technology and Society Magazine* 30, no. 3 (2011): 34–41.

60. Ann Logue, "Access Controlled: Limiting Employee Tracking," *RFID Journal*, February 6, 2006, https://www.rfidjournal.com/articles/view?2134.

61. John Edwards, "The Great Recession Spurs RFID Adoption in Europe," *RFID Journal*, November 1, 2010, www.rfidjournal.com/articles/view?7935.

62. Will Weissert, "Microchips Implanted in Mexican Officials," NBC News, July 14, 2004.

63. Logue, "Access Controlled," 1.

64. Gundars Kaupins and Robert Minch, "Legal and Ethical Implications of Employee Location Monitoring," *International Journal of Technology and Human Interaction* 2, no. 3 (2006): 16–35.

65. "Triax Technologies Powers Lettire Construction's Connected Jobsite with IoT Wearable Technology," Spot-r, 2017, https://www.triaxtec.com/workersafety/news/triax-technologies-powers-lettire-constructions-connected-jobsite-iot-wearable-technology/.

66. Kaupins and Minch, "Legal and Ethical Implications."

67. Chandana Unnithan, "Examining Innovation Translation of RFID Technology in Australian Hospitals through a Lens Informed by Actor-Network Theory" (Melbourne: Victoria University), http://vuir.vu.edu.au/25828/.

68. Unnithan, "Examining Innovation Translation," 33.

69. Unnithan, "Examining Innovation Translation," 248.

70. Jill A. Fisher and Torin Monahan, "Tracking the Social Dimensions of RFID Systems in Hospitals," *International Journal of Medical Informatics* 77, no. 3 (2008): 176–183.

71. Fisher and Monahan, "Tracking the Social Dimensions of RFID," 180.

72. Fisher and Monahan, "Tracking the Social Dimensions of RFID," 180.

73. Unnithan, "Examining Innovation Translation."

74. Emmeline Taylor, "Teaching Us to Be 'Smart'? The Use of RFID in Schools and Habituation of Young People to Everyday Surveillance," in *Surveillance Futures: Social and Ethical Implications of New Technologies for Children and Young People*, ed. Emmeline Taylor and Tonya Rooney (London: Routledge, 2017), 67.

75. Jim Morrison, "Someone to Watch Over You," *RFID Journal*, June 1, 2005, https://www.rfidjournal.com/articles/view?1743.

76. Kim Zetter, "School RFID Plan Gets an F," *Wired*, February 10, 2005, http://archive.wired.com/politics/security/news/2005/02/66554.

77. ACLU, "RFIDs in Mandatory Student ID Badges Violate Privacy Rights, Groups Charge," ACLU, 2005, https://www.aclu.org/news/parents-and-civil-liberties-groups-urge-northern-california-school-district-terminate-use; Mary Catherine O'Connor, "RFID Takes Attendance—and Heat," *RFID Journal*, February 16, 2005, http://www.rfidjournal.com/articles/view?1408.

78. "Concern at Pupil Data Microchips," BBC, November 23, 2007, http://news.bbc.co.uk/2/hi/uk_news/education/7110105.stm.

79. Wendy M. Grossman, "Is UK College's RFID Chip Tracking of Pupils an Invasion of Privacy?" *The Guardian*, November 19, 2013, https://www.theguardian.com/technology/2013/nov/19/college-rfid-chip-tracking-pupils-invasion-privacy.

80. Ekahau, "Case Study: Friedrich-von-Canitz School Safety Solution," Ekahau.com, 2013, http://www.ekahau.com/userData/ekahau/documents/case-studies/Friedrich-von-Canitz_school_safety_case_study.pdf;

Ekahau, "Idaho's Skyview High School Wins RFID Journal's Innovation Award for Its RFID Panic Buttons That Ensure School Safety—Blog—Real-Time Location System," Ekahau.com, 2014, http://www.ekahau.com/real-time-location-system/blog/2014/04/22/idahos-skyview-high-school-wins-rfid-journals-innovation-award-for-its-rfid-panic-buttons-that-ensure-school-safety/.

81. David Kravets, "Student Suspended for Refusing to Wear RFID Chip Returns to School," *Wired*, August 22, 2013, http://www.wired.com/2013/08/student-rfid-chip-flap/.

82. Benjamin Shmueli and Ayelet Blecher-Prigat, "Privacy for Children," *Columbia Human Rights Law Review* 42 (2011): 760.

83. Shmueli and Blecher-Prigat, "Privacy for Children," 760.

84. danah boyd, "Taken out of Context: American Teen Sociality in Networked Publics" (Cambridge, MA: MIT Press, 2008), http://www.danah.org/papers/TakenOutOfContext.pdf; H. Jenkins, "'Complete Freedom of Movement': Video Games as Gendered Play Spaces," in *The Game Design Reader: A Rules of Play Anthology*, ed. Katie Salen Tekinbas and Eric Zimmerman (Cambridge, MA: MIT Press, 2006), 330.

85. Emmeline Taylor, "Teaching Us to Be 'Smart'?"

86. Rebecca Jeschke, "Reading, Writing, and RFID Chips: A Scary Back-to-School Future in California," Electronic Frontier Foundation, 2010, https://www.eff.org/deeplinks/2010/08/reading-writing-and-rfid-chips-scary-back-school, n.p.

87. Olga Kharif, Jordan Robertson, and Rea Tan, "Biometric Passports Born in Malaysia No Cure for Human Error," Bloomberg.com, March 11, 2014, http://www.bloomberg.com/news/articles/2014-03-11/epassports-born-in-malaysia-no-cure-for-human-error-correct.

88. For a comprehensive guide to European Union passport procedures, see Frontex "Operational and Technical Security of Electronic Passports" (Warsaw: European Agency for the Management of Operational Cooperation at the External Borders of the Member States of the European

Union, 2011), http://frontex.europa.eu/assets/Publications/Research/Operational_and_Technical_Security_of_Electronic_Pasports.pdf.

89. You can go here to this page to see a full list of requirements for the US Visa Waiver program, which include the requirement for a machine-readable passport that has a digital chip with biometric information: https://www.dhs.gov/visa-waiver-program-requirements.

90. Mohammad A. Hossain, "RFID in National ID Cards: A Privacy Concern" (5th Asia Pacific Computing & Philosophy Conference, Tokyo, Japan, 2009).

91. Hossain, "RFID in National ID Cards"; Juels, Molnar, and Wagner, "Security and Privacy Issues in E-Passports," *First International Conference on Security and Privacy for Emerging Areas in Communications Networks (SECURECOMM '05)*, September 5–9, 2005, https://eprint.iacr.org/2005/095.pdf.

92. "An Open Letter to the ICAO," ACLU, 2004, https://www.aclu.org/sites/default/files/FilesPDFs/icaoletter.pdf.

93. Not all enhanced ID cards rely on this standard. For example, many enhanced ID cards, including passport cards in the United States, use EPCglobal Gen2 standard UHF RFID rather than the HF RFID found in passports.

94. Juels, Molnar, and Wagner, "Security and Privacy Issues in E-Passports."

95. Sean Fallon, "E-Passports Can Be Hacked and Cloned in Minutes," Gizmodo, 2008, http://gizmodo.com/5033923/e-passports-can-be-hacked-and-cloned-in-minutes.

96. Ulrich Beck, *Risk Society: Towards a New Modernity* (London: Sage, 1992).

97. Gerard Hancke, "Practical Attacks on Proximity Identification Systems," *Proceedings of IEEE Symposium on Security and Privacy* (Oakland, CA, 2006).

98. The short read range the researchers found is even more notable because Canadian enhanced ID cards use UHF RFID rather than the HF RFID found in passports.

99. Karen Louise Smith et al., "Playing with Surveillance: The Design of a Mock RFID-Based Identification Infrastructure for Public Engagement," *Surveillance & Society* 9, no. 1/2 (2011): 149–166.

100. David Lyon, "Surveillance, Security and Social Sorting: Emerging Research Priorities," *International Criminal Justice Review* 17, no. 3 (2007): 161–170.

101. Katja Franko Aas, "'The Body Does Not Lie': Identity, Risk and Trust in Technoculture," *Crime, Media, Culture* 2, no. 2 (2006): 143–158; David Lyon, "Surveillance, Security and Social Sorting: Emerging Research Priorities," *International Criminal Justice Review* 17, no. 3 (2007): 161–170.

102. Martha Kelner, "Call for Athletes to Be Fitted with Microchips in Fight against Drug Cheats," *The Guardian*, October 10, 2017, http://www.theguardian.com/sport/2017/oct/10/call-for-athletes-to-be-fitted-with-microchips-fight-against-drug-cheats.

103. The CAMCAT website can be found at http://camcat.org/.

104. M. G. Michael and Katina Michael, eds., *Uberveillance and the Social Implications of Microchip Implants: Emerging Technologies* (New York: Information Science Reference, 2014).

105. Torin Monahan and Jill A. Fisher, "Implanting Inequality: Empirical Evidence of Social and Ethical Risks of Implantable Radio-Frequency Identification (RFID) Devices," *International Journal of Technology Assessment in Health Care* 26, no. 4 (2010): 370–376.

106. Christine Perakslis, "Willingness to Adopt RFID Implants: Do Personality Factors Play a Role in the Acceptance of Uberveillance?" in *Uberveillance and the Social Implications of Microchip Implantation* (New York: Information Science Reference, 2014), 136.

107. James Katz and Ronald Rice, "Public Views of Mobile Medical Devices and Services: A US National Survey of Consumer Sentiments towards RFID Healthcare Technology," *International Journal of Medical Informatics* 78 (2009): 180.

108. Monahan and Fisher, "Implanting Inequality," 371.

109. Kenneth R. Foster and Jan Jaeger, "Ethical Implications of Implantable Radiofrequency Identification (RFID) Tags in Humans," *American Journal of Bioethics* 8, no. 8 (2008): 44–48; Mark Levine et al., "What Are the Benefits and Risks of Fitting Patients with Radiofrequency Identification Devices?" *PLoS Med* 5, no. 2 (2007).

110. Monahan and Fisher, "Implanting Inequality," 374.

111. Borut Werber and Anja Žnidaršič, "The Use of Subcutaneous RFID Microchip in Health Care—a Willingness to Challenge," *Health and Technology* 5, no. 1 (2015), doi:10.1007/s12553-015-0105-357.

112. Çigdem Benam, "Emergence of a 'Big Brother' in Europe: Border Control and Securitization of Migration," *Insight Turkey* 13, no. 3 (2011): 191–297.

113. Julie Boesen, Jennifer A. Rode, and Clara Mancini, "The Domestic Panopticon: Location Tracking in Families," in *Proceedings of the 12th ACM International Conference on Ubiquitous Computing*, Ubicomp '10 (New York: ACM, 2010), 65–74; Russell Spears and Martin Lea, "Panacea or Panopticon? The Hidden Power in Computer-Mediated Communication," *Communication Research* 21, no. 4 (1994): 427–459.

114. Kevin D. Haggerty, "Tear Down the Walls: On Demolishing the Panopticon," in *Theorizing Surveillance: The Panopticon and Beyond*, ed. David Lyon (Portland: Willan Publishing, 2006), 23–44; David Lyon, *Theorizing Surveillance: The Panopticon and Beyond* (Portland: Willan Publishing, 2006).

115. Daniel Solove, *Understanding Privacy* (Cambridge: Harvard University Press, 2008).

116. Edgar A. Whitley, Uri Gal, and Annemette Kjaergaard, "Who Do You Think You Are? A Review of the Complex Interplay between Information Systems, Identification and Identity," *European Journal of Information Systems* 23, no. 1 (2014): 17–35, esp. 21.

117. George Lakoff and Mark Johnson, *Metaphors We Live By* (Chicago: University of Chicago Press, 1980).

118. Turow, *The Daily You.*

119. Glenn Greenwald, "NSA Collecting Phone Records of Millions of Verizon Customers Daily," *The Guardian*, June 6, 2013, http://www.theguardian.com/world/2013/jun/06/nsa-phone-records-verizon-court-order.

120. Patricia A. Norberg, Daniel R. Horne, and David A. Horne, "The Privacy Paradox: Personal Information Disclosure Intentions versus Behaviors," *Journal of Consumer Affairs* 41, no. 1 (2007): 100–126.

121. The authors of *Spychips* are explicitly *against* governmental regulations of RFID. They take a more libertarian approach and argue that it should be up to consumers to fight against RFID.

Conclusion: The Future of Identification Infrastructures

1. William Gibson, "The Science in Science Fiction," NPR, November 30, 1999, https://www.npr.org/templates/story/story.php?storyId=1067220.

2. Will Oremus, "How 'Big Data' Went Bust," *Slate*, October 16, 2017, http://www.slate.com/articles/technology/technology/2017/10/what_happened_to_big_data.html.

3. Ben Lovejoy, "U.S. iPhone Market Share up on Last Year, but Samsung Takes #1 Slot," 9to5Mac, August 9, 2017, https://9to5mac.com/2017/08/09/us-iphone-sales-ios-market-share-kantar/.

4. Rian Boden, "Barcelona Introduces Contactless and NFC Mobile Payments across Public Transport Network," NFC World, November 13, 2015, https://www.nfcworld.com/2015/11/13/339571/barcelona-introduces-contactless-and-nfc-mobile-payments-across-public-transport-network/.

5. Henry Jenkins, *Convergence Culture: When Old and New Media Collide* (New York: New York University Press, 2006).

6. Bruce Sterling, *Shaping Things* (Cambridge, MA: MIT Press, 2005).

7. Bleeker, "A Manifesto for Networked Objects-Cohabiting with Pigeons, Arphids and Aibos in the Internet of Things," Near Future Laboratory, blog, February 26, 2006.

8. Mirca Madianou, "Smartphones as Polymedia," *Journal of Computer-Mediated Communication* 19, no. 3 (2014): 667–680.

9. Mark Roberti, "Apparel Edges toward the Tipping Point," *RFID Journal*, January 6, 2015, https://www.rfidjournal.com/articles/view?12566.

10. Roberti, "Apparel Edges toward the Tipping Point."

11. Swedberg, "RFID Brings Lululemon's Inventory Accuracy to 98 Percent," *RFID Journal*, April 18, 2016, http://www.rfidjournal.com/articles/view?14354.

12. Adrian Beck, "Measuring the Impact: Key Lessons from 10 Retailers Using RFID" (London: GS1 UK, 2018), ecr.pl/wp-content/uploads/2018/03/Measuring-the-Impact-of-RFID-in-Retailing.pdf.

13. The ten companies that participated in the study were Adidas, C&A, Decathlon, Jack Wills, John Lewis, Lululemon, Marc O'Polo, Marks & Spencer, River Island, and Tesco.

14. John T. Dunlop and Jan Rifkin, "Introduction," in *Revolution at the Checkout Counter* (Cambridge, MA: Harvard University Press, 1997), 13.

15. Claire Swedberg, "Sales Are Up and Overstocking Is Down, Study Reports, Due to RFID Use in Stores," *RFID Journal*, February 27, 2018, http://www.rfidjournal.com/articles/view?17286.

16. Lisa Manthei, "4 Important Differences between Multi-Channel and Omnichannel Marketing," Emarsys, 2016, https://www.emarsys.com/en/resources/blog/multi-channel-marketing-omnichannel/, n.p.

17. Swedberg, "Sales Are Up."

18. Claire Swedberg, "New Chip Features Privacy Function for European Retailers," *RFID Journal*, October 3, 2017, http://www.rfidjournal.com/articles/view?16676.

19. Claire Swedberg, "German Clothing Retailer Adler Gives RFID Robots a Spin," *RFID Journal*, February 12, 2016, http://www.rfidjournal.com/articles/view?14057/, n.p.

20. Swedberg, "German Clothing Retailer Adler," n.p.

21. Evan Schuman, "Will Tesco Shoppers Freak Out at Six-Foot Tall RFID Robots?" Computer World, June 8, 2015, https://www.computerworld.com/article/2932559/retail-it/will-tesco-shoppers-freak-out-at-six-foot-tall-rfid-robots.html.

22. Dani Deahl, "MIT Figured Out a Better Way for Drones to Use RFID Technology," The Verge, September 2, 2017, https://www.theverge.com/2017/9/2/16217302/mit-drones-billion-dollar-problem-inventory-rfid-tags.

23. Jennifer Wang et al., "A New Vision for Smart Objects and the Internet of Things: Mobile Robots and Long-Range UHF RFID Sensor Tags," Arxiv.org, 2015, https://arxiv.org/pdf/1507.02373.pdf.

24. Samar Warsi, "Inventory Managers Are Being Replaced by RFID-Mounted Drones," *Motherboard*, April 25, 2016, https://motherboard.vice.com/en_us/article/9a3kkz/inventory-managers-are-being-replaced-by-rfid-mounted-drones-AGE-Steel, n.p.

25. Geoffrey Bowker and Susan Leigh Star, *Sorting Things Out* (Cambridge, MA: MIT Press, 1999).

26. Nigel Thrift, "Remembering the Technological Unconscious by Foregrounding Knowledges of Position," *Environment and Planning D: Society and Space* 22 (2004): 175–190.

27. Jane Bennett, *Vibrant Matter: A Political Ecology of Things* (Durham, NC: Duke University Press, 2010).

28. John Durham Peters, *The Marvelous Clouds: Toward a Philosophy of Elemental Media* (Chicago: University of Chicago Press, 2015), 31.

Index

Note: Page numbers in italics refer to images; those followed by n refer to notes, with note number.

Printed in the United States
by Baker & Taylor Publisher Services